Adaptive Filtering - Theories and Applications

Adaptive Filtering - Theories and Applications

Editor

Lefteris Tyler

Adaptive Filtering - Theories and Applications

Edited by **Leftoris Tyler**

ISBN: 978-1-68117-207-1
Library of Congress Control Number: 2016934754

© 2017 by
SCITUS Academics LLC,
www.scitusacademics.com
Box No. 4766, 616 Corporate Way,
Suite 2, Valley Cottage,
NY 10989

Preface

An adaptive filter is a computational device that iteratively models the relationship between the input and output signals of the filter. An adaptive filter self-adjusts the filter coefficients according to an adaptive algorithm. Over the past three decades, digital signal processors have made great advances in increasing speed and complexity, and reducing power consumption. As a result, real-time adaptive filtering algorithms are quickly becoming practical and essential for the future of communications, both wired and wireless. An adaptive filter designs itself based on the characteristics of the input signal to the filter and a signal that represents the desired behaviour of the filter on its input. Because of the complexity of the optimization algorithms, almost all adaptive filters are digital filters. Adaptive filters are required for some applications because some parameters of the desired processing operation are not known in advance or are changing. The closed loop adaptive filter uses feedback in the form of an error signal to refine its transfer function. Adaptive filtering can be used to characterize unknown systems in time-variant environments. Commonly, the closed loop adaptive process involves the use of a cost function, which is a criterion for optimum performance of the filter, to feed an algorithm, which determines how to modify filter transfer function to minimize the cost on the next iteration. The most common cost function is the mean square of the error signal. This book, Adaptive Filtering - Theories and Applications, offers some theoretical approaches and practical applications in diverse areas that support increasing of adaptive systems. The book reflect the latest advances in this field; particularly an increased coverage given to the practical applications of the theory to illustrate the much broader range of adaptive filters applications developed in recent years.

TABLE OF CONTENTS

CHAPTER 1 . Applications of Adaptive Filtering...................................... **1**

CHAPTER 2 A Robust Weighted Combination Forecasting Method Based on Forecast Model Filtering and Adaptive Variable Weight Determination.. **25**

CHAPTER 3 Gas Turbine Transient Performance Tracking Using Data Fusion Based on an Adaptive Particle Filter **61**

CHAPTER 4 Robust Hammerstein Adaptive Filtering under Maximum Correntropy Criterion **87**

CHAPTER 5 Application of Adaptive Noise Cancellation in Transabdominal Fetal Heart Rate Detection Using Photoplethysmography. **111**

CHAPTER 6 Noise Removal from EEG Signals in Polisomnographic Records Applying Adaptive Filters in Cascade **133**

CHAPTER 7 Adaptive Filtering by Non-Invasive Vital Signals Monitoring and Diseases Diagnosis... **163**

CHAPTER 8 Adaptive Filters for Processing Water Level Data................ **185**

CHAPTER 9 Anti-Multipath Filter with Multiple Correlators in GNSS Receviers **205**

CHAPTER 10 Adaptive Data Filtering of Inertial Sensors with Variable Bandwidth **229**

CHAPTER 11 Noise Reduction and Gap Filling of fAPAR Time Series Using an Adapted Local Regression Filter **251**

CHAPTER 12 .. VDTA-Based Wave Active Filter **279**

Index **291**

Chapter 1

Applications of Adaptive Filtering

J. Gerardo Avalos[1], Juan C. Sanchez and Jose Velazquez

[1] National Polytechnic Institute, Mexico

1. INTRODUCTION

Owing to the powerful digital signal processors and the development of advanced adaptive algorithms there are a great number of different applications in which adaptive filters are used. The number of different applications in which adaptive techniques are being successfully used has increased enormously during the lasttwo decades.There is a wide variety of configurations that could be applied in different fields such telecommunications, radar, sonar, video and audio signal processing, noise reduction, between others.

The efficiency of the adaptive filters mainly depends on the design technique used and the algorithm ofadaptation. The adaptive filters can be analogical designs, digital or mixed which show their advantages and disadvantages, for example, the analogical filters are low power consuming and fast response, but they represent offset problems, which affect the operation of the adaptation algorithm (Shoval et al., 1995). The digital filters are offset free and offeran answer of greater precision. Also the adaptive filters can be a combination of different types of filters, like single-input or multi-input filters, linear or nonlinear, and finite impulse response FIR or infinite impulse response IIR filters.

The adaptation of the filter parameters is based on minimizing the mean squared error between the filter output and a desired signal.The most common adaptation algorithms are, Recursive Least Square (RLS), and the Least Mean Square (LMS), where RLS algorithm offers a higher convergence speed compared to the LMS algorithm, but as for computation complexity, the LMS algorithm maintains its advantage. Due to the computational simplicity, the LMS algorithm is most commonly used in the design andimplementation of integrated adaptive filters.The LMS digital algorithm is based on the gradientsearch according to the equation (1).

$$w(n + 1) - w(n) + \mu e(n)x(n) \tag{1}$$

Where w(n) is the weights vector in the instant n, w(n+1) is equal to the weights vector in n+1, x(n) is the input signal simple vector which is stored in the filter delayed line, where e(n) corresponds to the filter's error, which is the difference between the desired signal and the output filter's signal, and μ is the filter's convergence factor. The convergence factor μ determines the minimum square average error and the convergence speed. This factor is directly proportional to the convergence speed and indirectly proportional to the minimal error. Then a convergence speed and minimal error relation is established.

The application depends on the adaptive filter configuration used. The classical configurations of adaptive filtering are system identification, prediction, noise cancellation, and inverse modeling. The differences between the configurations are given by the way the input, the desired and the output signals are used.The main objective of this chapter is to explainthe typical configurations and it will focus on recent applications of adaptive filtering that are used in the real world.

2. SYSTEM IDENTIFICATION

The system identification is an approach to model an unknown system. In this configuration the unknown system is in parallel with an adaptive filter, and both are excited with the same signal. When the output MSE is minimized the filter represents the desired model.

The structure used for adaptive system identification is illustrated in Figure 1, where P(z) is an unknown system to be identified by an adaptive filter W(z). The signal x(n) excites P(z) and W(z), the desired signal d(n) is the unknown system output, minimizing the difference of output signals y(n) and d(n), the characteristics of P(z) can be determined.

The estimation error is given as (2)

$$e(n) = d(n) - y(n) = \sum_{l=0}^{L-1} [p(l) - w_1(n)]x(n-l) \tag{2}$$

Where p(l) is the impulse respond of the unknown plant, By choosing each $w_1(n)$ close to each p(l), the error will be minimized. For using white noise as the excitation signal, minimizing e(n) will force the $w_1(n)$ to approach p(l), that is,

$$w_1(n) \approx p(l),\ l = 0,\ 1,\dots,\ L-1 \qquad\qquad (3)$$

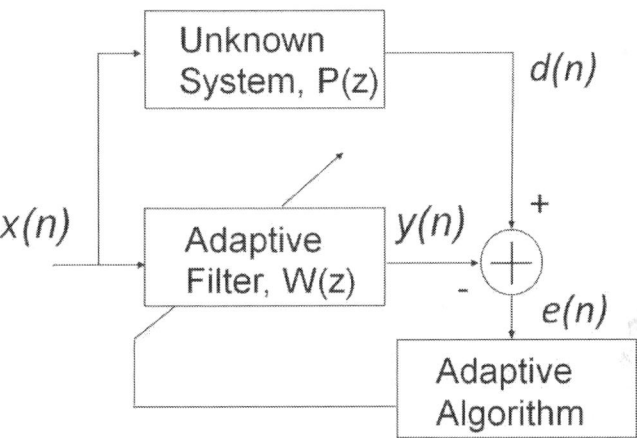

Figure 1. Adaptive filter for system identification

When the difference between the physical system response d(n) and the adaptive model response y(n) has been minimized, the adaptive model approximates P(z) from the input/output viewpoint. When the plan is time varying, the adaptive algorithm has the task of keeping the modelling error small by continually tracking time variations of the plant dynamics.

Usually, the input signal is awideband signal, in order to allow the adaptive filter to converge to a good model of the unknownsystem.If the input signal is a white noise, the best model for the unknown system is a system whose impulseresponse coincides with the N + 1 first samples of the unknown system impulse response. In thecases where the impulse response of the unknown system is of finite length and the adaptive filteris of sufficient order, the MSE becomes zero if thereis no measurement noise (or channel noise). In practical applications the measurement noise isunavoidable, and if it is uncorrelated with the input signal, the expected value of the adaptive-filtercoefficients will coincide with the unknown-system impulse response samples. The output errorwill of course be the measurement noise (Diniz, 2008). Some real world applications of the system identification scheme include control systems and seismic exploration.

3. LINEAR PREDICTOR

The linear prediction estimates the values of a signal at a future time. This model is wide usually in speech processing applications such as speech coding in cellular telephony, speech enhancement, and speech recognition. In this configuration the desired signal is a forward version of the adaptive filter input signal. When the adaptive algorithm convergences the filter represents a model for the input signal, this model can be used as a prediction model. The linear prediction system is shown in Figure 2.

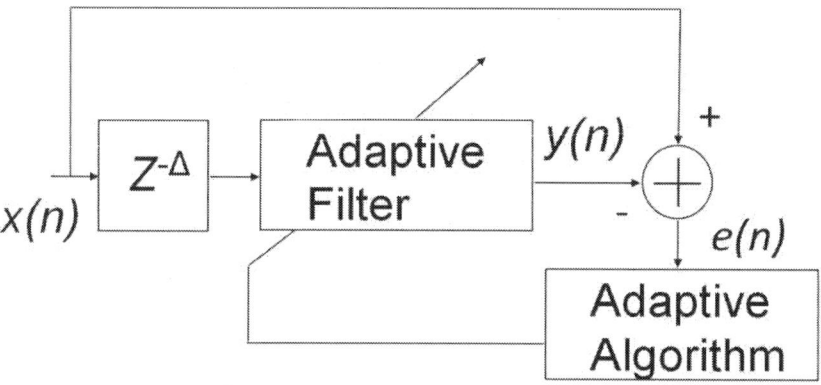

Figure 2. Adaptive filter for linear prediction

The predictor output y(n) is expressed as

$$y(n) = \sum_{l=0}^{L-1} w_1(n)x(n - \Delta - l) \qquad (4)$$

Where Δ is the number of delay samples, so if we are using the LMS algorithm the coefficients are updated as

$$w(n + 1) = w(n) + \mu x(n - \Delta)e(n) \qquad (5)$$

Where x(n - Δ) = [x(n - Δ) x(n - Δ -1)... x(n - Δ - L + l)]^T is then delayed reference signal vector, and e(n) = x(n) − y(n) is the prediction error. Proper selection of the prediction delay Δ allows improved frequency estimation performance for multiple sinusoids in white noise.

Atypical predictor's application is in linear prediction coding of speech signals, where the predictor'stask is to estimate the speech parameters. These parameters are part of the coding informationthat is transmitted or stored along with other information inherent to the speech characteristics, suchas pitch period, among others.

The adaptive signal predictor is also used for adaptive line enhancement (ALE), where the input signalis a narrowband signal (predictable) added to a wideband signal. After convergence, the predictoroutput will be an enhanced version of the narrowband signal.Yet another application of the signal predictor is the suppression of narrowband interference in awideband signal. The input signal, in this case, has the same general characteristics of the ALE.

4. INVERSE MODELING

The inverse modeling is an application that can be used in the area of channel equalization, for example it is applied in modems to reduce channel distortion resulting from the high speed of data transmission over telephone channels. In order to compensate the channel distortion we need to use an equalizer, which is the inverse of the channel's transfer function.

High-speed data transmission through channels with severe distortion can be achieved in several ways, one way is to design the transmit and receive filters so that the combination of filters and channel results in an acceptable error from the combination of intersymbol interference and noise; and the other way is designing an equalizer in the receiver that counteracts the channel distortion. The second method is the most commonly used technology for data transmission applications.

Figure 3 shows an adaptive channel equalizer, the received signal y(n) is different from the original signal x(n) because it was distorted by the overall channel transfer function C(z), which includes the transmit filter, the transmission medium, and the receive filter.

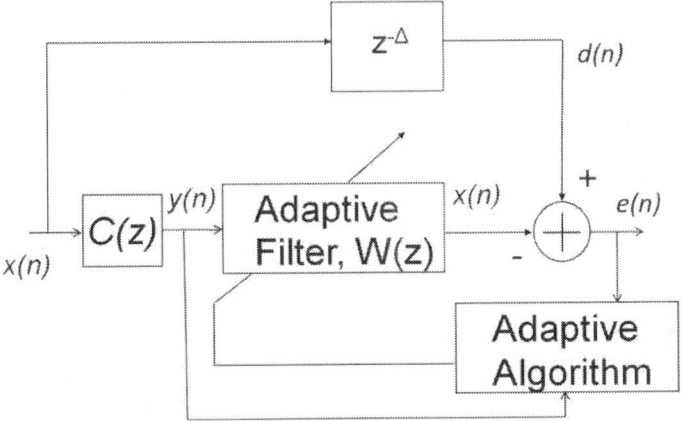

Figure 3. Adaptive Channel equalizer.

To recover the original signal x(n), y(n) must be processed using the equalizer W(z), which is the inverse of the channel's transfer function C(z) in order to compensate for the channel distortion. Therefore the equalizer must be designed by

$$W(z) = \frac{1}{C(z)} \tag{6}$$

In practice, the telephone channel is time varying and is unknown in the design stage due to variations in the transmission medium. Thus it is needed an adaptive equalizer that provides precise compensation over the time-varying channel. The adaptive filter requires the desired signal d(n) for computing the error signal e(n) for the LMS algorithm. An adaptive filter requires the desired signal d(n) for computing the error signal e(n) for the LMS algorithm.

The delayed version of the transmitted signal x(n - Δ) is the desired response for the adaptive equalizer W(z). Since the adaptive filter is located in the receiver, the desired signal generated by the transmitter is not available at the receiver. The desired signal may be generated locally in the receiver using two methods. During the training stage, the adaptive equalizer coefficients are adjusted by transmitting a short training sequence. This known transmitted sequence is also generated in the receiver and is used as the desired signal d(n) for the LMS algorithm.

After the short training period, the transmitter begins to transmit the data sequence. In the data mode, the output of the equalizer x(n) is used by a decision device to produce binary data. Assuming that the output of the decision device is correct, the binary sequence can be used as the desired signal d(n) to generate the error signal for the LMS algorithm.

5. JAMMER SUPPRESSION

Adaptive filtering can be a powerful tool for the rejection of narrowband interference in a direct sequence spread spectrum receiver. Figure 4 illustrates a jammer suppression system. In this case the output of the filter y(n), is an estimate of the jammer, this signal is subtracted from the received signal x(n), to yield an estimate of the spread spectrum.

To enhance the performance of the system a two-stage jammer suppressor is used. The adaptive line enhancer, which is essentially another adaptive filter, counteracts the effects of finite correlation which leads to partial cancellation of the desired signal. The number of coefficients required for either filter is moderate, but the sampling frequency may be well over 400 KHz.

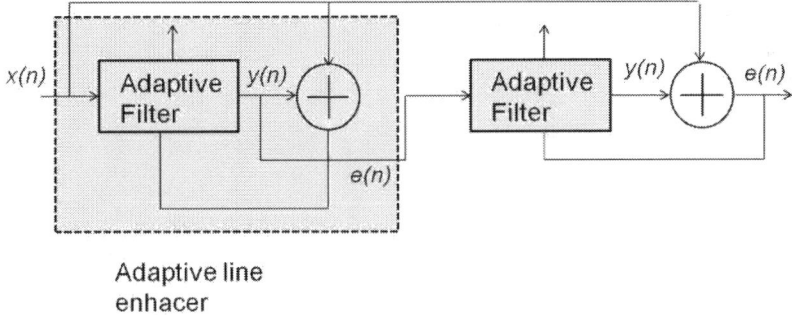

Adaptive line
enhacer

Figure 4. Jammer suppression in direct sequence spread spectrum receiver

6. ADAPTIVE NOTCH FILTER

In certain situations, the primary input is a broadband signal corrupted by undesired narrowband (sinusoidal)interference. The conventional method of eliminating such sinusoidal interference is using anotch filter that is tuned to the frequency of the interference (Kuo et al., 2006). To design the filter, we need the precisefrequency of the interference. The adaptive notch filter has the capability to track the frequency of theinterference, and thus is especially useful when the interfering sinusoid drifts in frequency.A single-frequency adaptive notch filter with two adaptive weights is illustrated in Figure 5,where the input signal is a cosine signal as

$$x(n) = x_0(n)Acos(\omega_o n) \qquad (7)$$

A 90 phase shifter is used to produce the quadrature signal

$$x_1(n) = Asin(\omega_0 n) \qquad (8)$$

For a sinusoidal signal, two filter coefficients are needed. The reference input is used to estimate the composite sinusoidal interfering signal contained in theprimary input d(n). The center frequency of the notch filter is equal to the frequency of the primary sinusoidal noise. Therefore, the noise at that frequency is attenuated. This adaptive notch filter providesa simple method for eliminating sinusoidal interference.

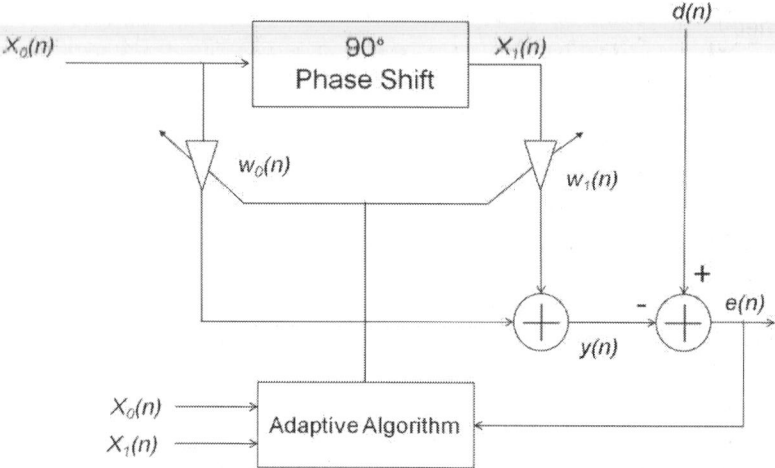

Figure 5. Adaptive Notch Filter

7. NOISE CANCELLER

The noise cancellers are used to eliminate intense background noise. This configuration is applied in mobile phones and radio communications, because in some situations these devices are used in high-noise environments. Figure 6 shows an adaptive noise cancellation system.

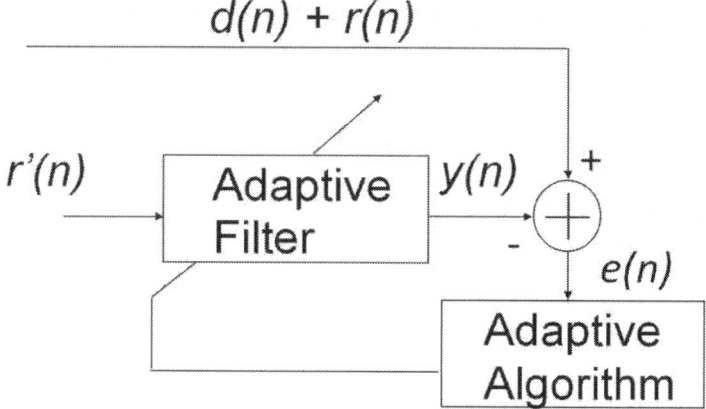

Figure 6. Adaptive noise canceller system

The canceller employs a directional microphone to measure and estimate the instantaneous amplitude of ambient noise r'(n), and another microphone is used to take the speech signal which is contaminated with noise d(n) + r(n). The ambient noise is processed by the adaptive filter to make it equal to the noise contaminating the speech signal, and then is subtracted to cancel out the noise in the desired signal. In order to be effectively the ambient noise must be highly correlated with the noise components in the speech signal, if there is no access to the instantaneous value of the contaminating signal, the noise cannot be cancelled out, but it can be reduced using the statistics of the signal and the noise process. Figure 7 shows a voice signal with noise; those signals were used in noise canceller system implemented on a digital signal processor. The desired signal is a monaural audio signal with sampling frequency of 8 KHz. The noise signal is an undesired monaural musical piece with a sampling frequency of 11 KHz. As it can be seen in the image the desired signal is highly contaminated, so in this structure it must be used a fast adaptation algorithm in order to reach the convergence and eliminate all the unwanted components from the desired signal.

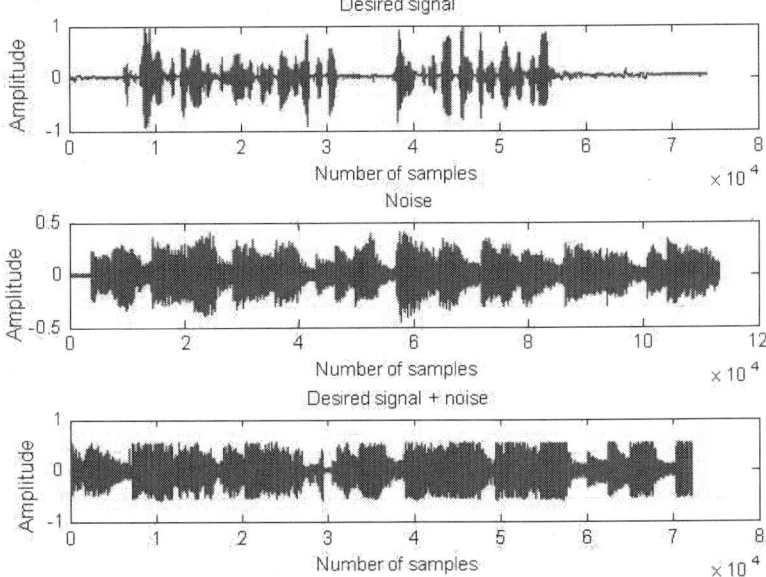

Figure 7. Signals used in the noise canceller system.

The frequency analysis of the signals used in the noise canceller system can be seen on the spectrograms of the Figure 8. The Figure shows that the output signal has some additionalfrequency components with respect to the input signal.

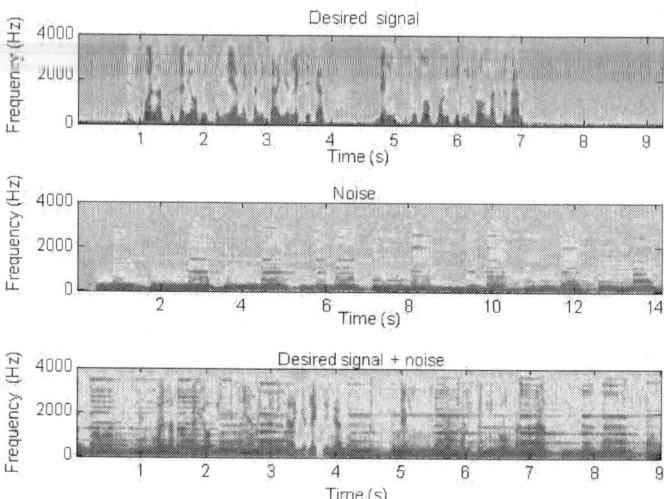

Figure 8. Spectrograms of the signals used in the noise canceller system

The output of the noise canceller is the error signal, the Figure 9 shows the error signal obtained when it is used an LMS algorithm. With the spectrogram of the signal it is shown that all the undesired frequency components were eliminated.

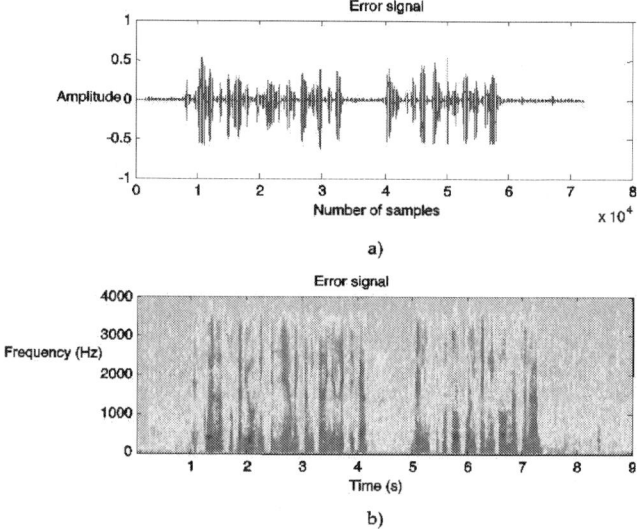

Figure 9. a) Time waveform of the output signal b) Spectrogram of the output signal.

The adaptive noise canceller system is used in many applications of active noise control (ANC), in aircrafts is used to cancel low-frequency noise inside vehicle cabins for passenger comfort. Most major aircraft manufacturers are developing such systems, mainly for noisy propeller-driven airplanes. In the automobile industry there are active noise cancellation systems designed to reduce road noise using microphones and speakers placed under the vehicle's seats.

Another application is active mufflers for engine exhaust pipes, which have been in use for a while on commercial compressors, generators, and such. With the price for ANC solutions dropping, even automotive manufacturers are now considering active mufflers as a replacement of the traditional baffled muffler for future production cars. The resultant reduction in engine back pressure is expected to result in a five to six percent decrease in fuel consumption for in-city driving.

Another application that has achieved widespread commercial success are active headphones to cancel low-frequency noise.The active headphones are equipped with microphones on outside of the ear cups that measure the noise arriving at the headphones. This noise is then being cancelled by sending the corresponding "anti-noise" to the headphones' speakers. For feedforward ANC, the unit also includes a microphone inside each ear cup to monitor the error - the part of the signal that has not been canceled by the speakers in order to optimize the ANC algorithm. Very popular with pilots, active headphones are considered essential in noisy helicopters and propeller-powered airplanes.

7.1. Echo Cancellation

In telecommunications, echo can severely affect the quality and intelligibility of voiceconversation in telephone, teleconference or cabin communication systems. The perceivedeffect of an echo depends on its amplitude and time delay. In general, echoes with appreciableamplitudes and a delay of more than 1 ms can be noticeable. Echo cancellation is an important aspect of the design of modern telecommunicationssystems such as conventional wire-line telephones, hands-free phones, cellular mobile(wireless) phones, teleconference systems and in-car cabin communication systems.

In transmission networks the echoes are generated when a delayed and attenuated version of the signal sent by the local emitter to the distant receiver reaches the local receiver. These echo signals have their origin in the hybrid transformers which perform the two/four-wire conversion, in the impedance mismatches along the two-wire lines, and in some cases in acoustic couplings between loudspeakers and microphones in the subscriber sets.

The echo cancellation consists in modelling these unwanted couplings between local emitters and receivers and subtracting a synthetic echo from the real echo. According to the nature of the signals involved, the system will work as echo data canceller or voice echo canceller.

7.1.2. Voice Echo Canceller

Due to the characteristics of the speech signal, the voice echo cancellation system is somewhat different from the data echo canceller. The speech is a high level nonstationary signal, and due to the signal bandwidth and the velocity of the acoustic waves in the open air, the filters must have a very long number of coefficients. Also in order to reach a high level of performance and meet the expectations of the user, the voice echo canceller may have several other functions, like speech detection and denoising.

Figure 10 illustrates the operation of an adaptive line echo canceller. The speech signal on the line from speaker A to speaker B is input to the four/two-wire hybrid B and to the echo canceller. The echo canceller monitors the signal on line from B to A and attempts tomodel the echo path and synthesise a replica of the echo of speaker A. This replica is used to subtract and cancel out the echo of speaker A on the line from B to A. The echo canceller is basically an adaptive linear filter. The coefficients of the filter are adapted so that the energy of the signal on the line is minimised.

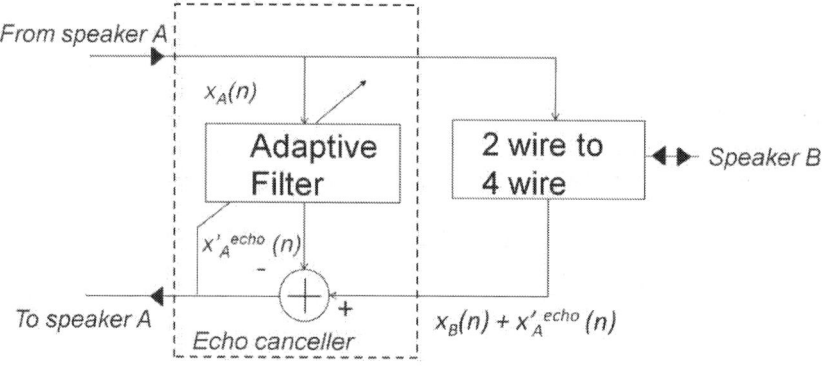

Figure 10. Adaptive echo cancellation system.

Assuming that the signal of the line from speaker B to speaker A, $y_B(n)$, is composed of the speech of speaker B, $x_B(n)$, plus the echo of speaker A, $x_A^{echo}(n)$,

$$y_B(n) = x_B(n) + x_A^{echo}(n) \tag{9}$$

Speech and echo signals are not simultaneously present on a phone line unless both speakers are speaking simultaneously. Assuming that the truncated impulse response of the echo path is modelled by an FIR filter, the output estimate of the synthesised echo signal can be expressed as

$$x'^{echo}_A(n) = \sum_{l=0}^{P} h_l(n) x_A(n-l) \qquad (10)$$

Where $h_l(n)$ are the time varying coefficients of an adaptive FIR filter model of the echo path and $x'_A{}^{echo}(n)$ is an estimate of the echo of speaker A on the line from speaker B to speaker A. The residual echo signal, or the error signal, after echo subtraction is given by

$$e(n) = y_B(n) - x'^{echo}_A(n) = x_B(n) + x^{echo}_A(n) - \sum_{l=0}^{P} h_l(n) x_A(n-l)$$

For those time instants when speaker A is talking and speaker B is listening and silent, and only echo is present from line B to A, we have

$$e(m) = x'^{echo}_A(n) = x^{echo}_A(n) - x'^{echo}_A(n) = x^{echo}_A(n) - \sum_{l=0}^{P} h_l(n) x_A(n-l)$$

Where $x'_A{}^{echo}(n)$ is the residual echo.

In some cases it may happen the double talk situation, in this case both users talk at the same time, and simultaneous bidirectional transmission takes place. In this way it could be produced misalignment of the coefficients and a drop in echo attenuation, one way to solve this problem is holding the coefficients during double talk, but for this it is needed a double-talk detector. The performance of double-talk detectors is crucial for the comfort of the users.

7.1.3. Data Echo Canceller

Echo cancellation becomes more complex with the increasing integration of wireline telephone systems and mobile cellular systems, and the use of digital transmission methods such as asynchronous transfer mode (ATM) for integrated transmission of data, image and voice.

Those systems use full-duplex transmission data signals that are transmitted simultaneously in two directions and in the same frequency bands, meanwhile in half-duplex transmission just one direction are used at a time. The Figure 11 shows the principle of full-duplex transmission. The signal xA(N) is sent from terminal A to terminal B through a two wire line. The signal y(n) at the input of the receiver of terminal A consists of two components, a signal from the terminal B (yB(n)), which is the useful data signal, and the returned unwanted echo generated from xA(n). H(z) is a filter that is going to generate a synthetic echo y'(n) as close as possible to xA(n), after subtraction, the output error e(n) is kept sufficiently close to yB(n) to make the transmission of data from terminal D to terminal A satisfactory.

The number of coefficients (N) of the adaptive filter is derived from the duration of the echo impulse response that has to be compensated, taking into account the sampling frequency. In order to calculate the number of coefficients we could use

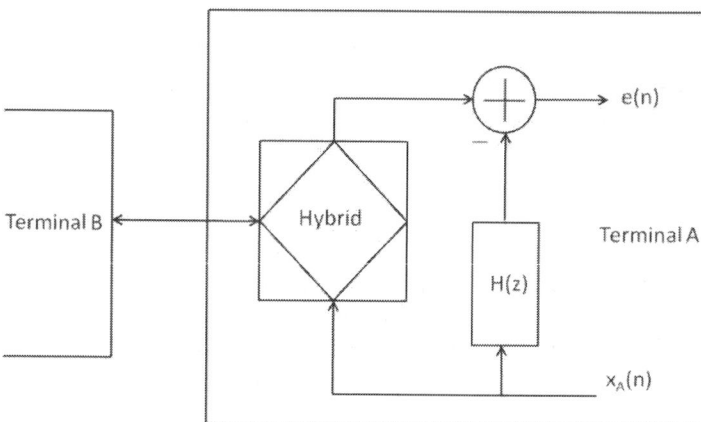

Figure 11. Echo cancellation for full-duplex transmission

$$N = (2D/v)\ fs \qquad\qquad (13)$$

Where N is the number of coefficients, D is the length of the line, v is the electrical signal velocity over the subscriber line and fs is the sampling frequency (Bellanger, 2001). Since the characteristics of the transmission line may change with time it is necessary to implement an adaptive filter.

7.1.4. Acoustic Echo

Acoustic echo results from a feedback path set up between the speaker and the microphone in a mobile phone, hands-free phone, teleconference or hearing aid system. Acoustic echo is reflected from a multitude of different surfaces, such as walls, ceilings and floors, and travels through different paths. If the time delay is not too long, then the acoustic echo may be perceived as a soft reverberation, and may add to the artistic quality of the sound; concert halls and church halls with desirable reverberation characteristics can enhance the quality of a musical performance.

Acoustic echo can result from a combination of direct acoustic coupling and multipath effect where the sound wave is reflected from various surfaces and then picked up by the microphone. In its worst case, acoustic feedback can result in howling if a significant proportion of the sound energy transmitted by the loudspeaker is received back at the microphone and circulated in the feedback loop.

The most effective method of acoustic feedback removal is the use of an adaptive feedback cancellation system (AFC). Fig. 12 illustrates a model of an acoustic feedback environment, comprising a microphone, a loudspeaker and the reverberating space of a room (Vaseghi, 2006). The z transfer function of a linear model of the acoustic feedback environment may be expressed as

$$H(z) = \frac{G(z)}{1 - G(z)A(z)} \tag{14}$$

Where G(z) is the z transfer function model for the microphone loudspeaker system and A(z) is the z transfer function model of reverberations and multipath reflections of a room environment. Assuming that the microphone loudspeaker combination has a flat frequency response with a gain G, the equation can be simplified to

$$H(z) = \frac{G}{1 - GA(z)} \tag{15}$$

Owing to the reverberation character of the room, the acoustic feedback path A(z) is itself a feedback system. The reverberating characteristics of the acoustic environment may be modelled by an all-pole linear predictive model, or alternatively a relatively long FIR model. The equivalent time-domain input/output relation for the linear filter model of equation (4) is given by the following difference equation

$$y(n) = \sum_{l=0}^{P} a_l(n)y(n - l) + Gx(n) \tag{16}$$

Where $a_l(n)$ is the coefficient of an all pole linear feedback model of the reverberating room environment, G is the microphone loudspeaker amplitude gain factor, and x(n) and y(n) are the time domain input and output signals of the microphone loudspeaker system.

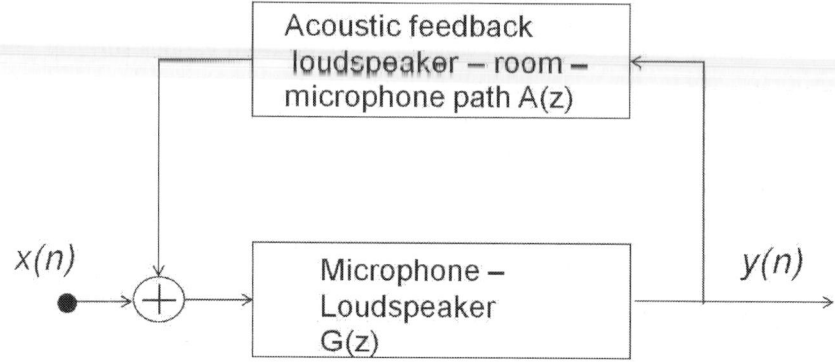

Figure 12. Acoustic feedback model

The most successful acoustic feedback control systems are based on adaptive estimation and cancellation of the feedback signal. As in a line echo canceller, an adaptive acoustic feedback canceller attempts to synthesise a replica of the acoustic feedback. The problem of acoustic echo cancellation is more complex than line echo cancellation for a number of reasons. First, acoustic echo is usually much longer (up to a second) than terrestrial telephone line echoes. In fact, the delay of an acoustic echo is similar to or more than a line echo routed via a geostationary satellite system. The large delay of an acoustic echo path implies that impractically large filters on the order of a few thousand coefficients may be required. An important application of acoustic feedback cancellation is in hearing aid systems.

7.1.5. Multiple-Input Multiple-Output (Mimo) Echo Cancellation
Multiple-input multiple-output (MIMO) echo-cancellation systems have applications in carcabin communications systems, stereophonic teleconferencing systems and conference halls.

Stereophonic echo cancellation systems have been developed relatively recently and MIMOsystems are still the subject of ongoing research and development.In a typical MIMO system there are P speakers and Q microphones in the room. As thereis an acoustic feedback path set up between each speaker and each microphone, there arealtogether P ×Q such acoustic feedback paths that need to be modelled and estimated. Thetruncated impulse response of each acoustic path from loudspeaker i to microphone $j$ismodelled by an FIR filter hij. The truncated impulse response of each acoustic path from ahuman speaker i to microphone j is modelled by an FIR filter, gij.For a large number of speakers and microphones, the modelling and identification ofthe numerous acoustic channels becomes a major problem due to the correlations of theecho signals, from a common number of sources, propagating through different channels, asdiscussed below.

7.2. Adaptive Feedback Cancellation In Hearing Aids

The hearing-aid processing amplifies the input signal to compensate for the hearing loss of the users. When this amplification is larger than the attenuation of the feedback path, instability occurs and usually results in feedback whistling, which limits the maximum gain that can be achieved.

Acoustic feedback in hearing aids refers to the acoustical coupling between the loudspeaker (also known as the receiver) and the microphone of the hearing aid. Because of this coupling, the hearing aid produces a severe distortion of the desired signal and an annoying howling sound when the gain is increase.

If the Feedback transfer function was known, it can be compensated for in the hardware, but the problem here is the time variability of the dynamics, caused by a change in interference characteristics. Some possible causes of this problem are hugs or objects like a telephone coming close to the ear.

There are several techniques to reduce the negative effects introduced by acoustic feedback. They can be broadly classified into feedforward suppression and feedback cancellation techniques. In feedforward suppression techniques, the regular signal processing path of the hearing aid is modified in such a way that it isstable in conjunction with the feedback path.

The most common technique is the use of a notch filter. In a notch filter, the gain is reduced in a narrow frequency band around the critical frequencies whenever feedback occurs. Nevertheless feedforward suppression techniques allcompromise the basic frequency response of the hearing aid, and, hence, may seriouslyaffect the sound quality (Spriet et al., 2006).A more promising solution for acoustic feedback is the use of a feedback cancellation system.

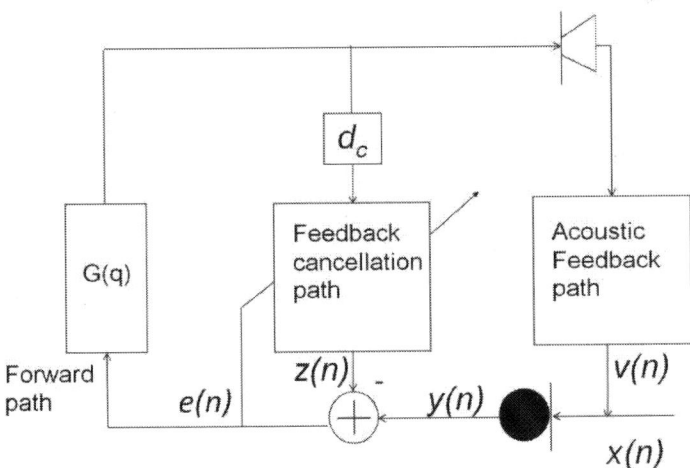

Figure 13. Adaptive feedback canceller

Figure 13 illustrates an adaptive feedback canceller, which produces an estimate z(n) of the feedback signal v(n) and subtracts this estimate z(n) from the microphone signal, so that, ideally, only the desired signal is preserved at the input of the forward path. Since the acoustic path between the loudspeaker and the microphone can vary significantly depending on the acoustical environment, the feedback canceller must be adaptive.

When the external input signal is correlated with the receiver input signal, the estimateof the feedback path is biased. This so-called "bias problem" results in a large modeling error and acancellation of the desired signal (Ma, 2010).

7.3. Foetal Monitoring, Cancelling of Maternal Ecg During Labour

Information derived from the foetal electrocardiogram (ECG), such as the foetal heart rate pattern, is valuable in assessing the condition of the baby before or during the childbirth. The ECG derived from electrodes placed on the mother's abdomen is susceptible to contamination from much larger background noise (for example muscle activity and foetal motion) and the mother's own ECG.

Considering the problem as an adaptive noise cancellation, where foetal ECG is a desired signal d(n), corrupted by the maternal signal r(n), a kind of additive noise. The measured foetal signal (MFECG(n)) from foetal lead can be expressed as

$$\text{MFECG}(n) = d(n) + r(n) \tag{17}$$

Another measurement MMECG(n) from maternal lead is given as a reference signal, that is correlated with r(n)and uncorrelated with d(n). MMECG can be used to estimate the noise r(n) by minimizing the mean square error. Figure 14 shows the block diagram for the enhancement of foetal ECG.

An adaptive filter is used to estimate maternal components in measured foetal ECG (MFECG) from measured maternal ECG (MMECG). The estimated components then are subtracted from the MFECG to obtain adaptive filtered foetal ECG (AFECG), in which maternal components are suppressed. Other artefacts, such as muscular contraction from maternal body and foetal movement, will induce baseline drift in the MFECG (Chen et al., 2000).

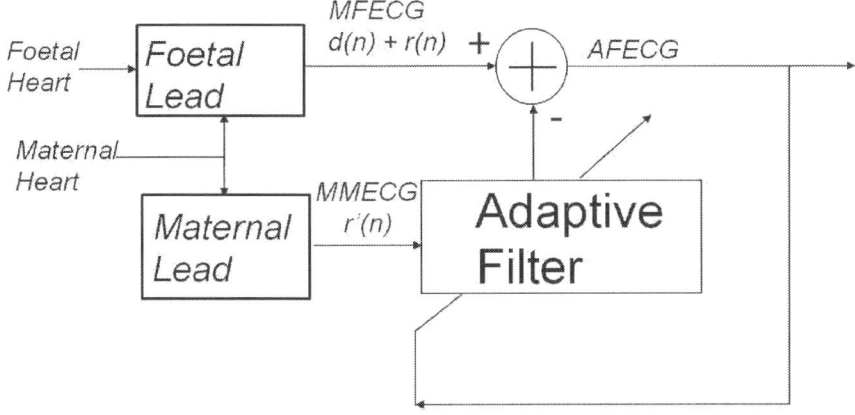

Figure 14. Adaptive cancelling of maternal ECG in foetal ECG

7.4. Removal of Ocular Artifacts From Electro-Encephalogram by Adaptive Filtering

The eye forms an electric dipole, where the cornea is positive and the retina is negative. When the eye moves (saccade, blink or other movements), the electric field around the eye changes, producing an electrical signal known as the electro-oculogram (EOG). As this signal propagates over the scalp, it appears in the recorded electro-encephalogram (EEG) as noise or artifacts that present serious problems in EEG interpretation and analysis. There are at least two kinds of EOG artifact to be removed: those produced by the vertical eye movement (the corresponding EOG is called VEOG) and those produced by the horizontal eye movement (HEOG). Consequently, a noise canceller with two reference inputs is used in this application (He et al., 2004).

Fig. 15 shows the EOG noise canceller. The primary input to the system is the EEG signal s(n), picked up by a particular electrode. This signal is modelled as a mixture of a true EEG x(n) and a noise component r(n). v(n) and v'(n) are the two reference inputs, VEOG and HEOG, respectively. v(n) and v'(n) are correlated, in some unknown way, with the noise component r(n) in the primary input. The desired output from the noise canceller e(n) is the corrected, or clean, EEG.

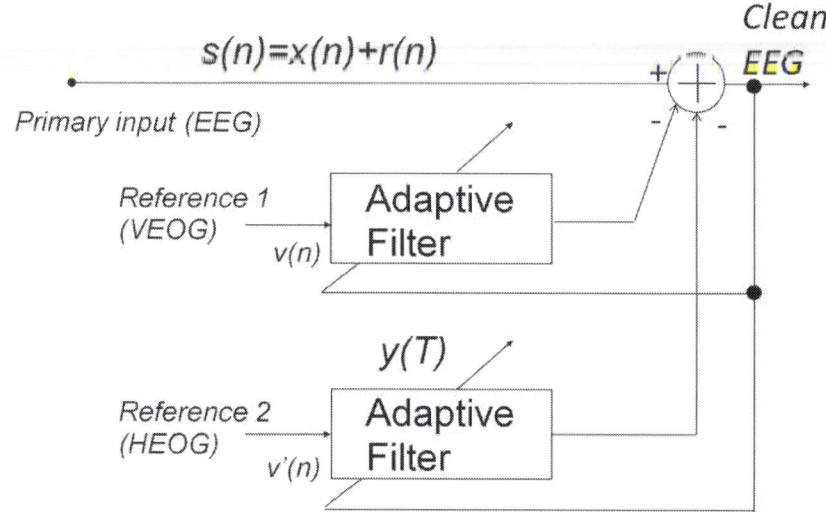

Figure 15. EOG noise canceller

7.5. Application of Adaptive Noise Cancelling Filters in AC Electricalmeasurements

Through adaptive noise cancellation it could be improved the ac electrical measurements. Often ac measurement circuits are influenced by noise caused by line frequency beat. The Figure 16 shows a system that cancels the line frequency beat. An ADC is used to sample the suitably divided down line voltage in order to determine the phase relative to the signal channel, which is sampled with a second ADC. The phase data is used as the noise input to an adaptive noise-cancelling filter used to cancel the effect on the transconductance amplifier output data (Wright et al., 2010).

Another common interference in ac measurement circuits is the coupling of the magnetic field generated by a nearby source. In such situations it may be possible to use an adaptive interference cancelling system with a simple coil system to measure the ambient magnetic field that causes the unwanted interference and then remove this interference from data obtained from a measurement circuit.

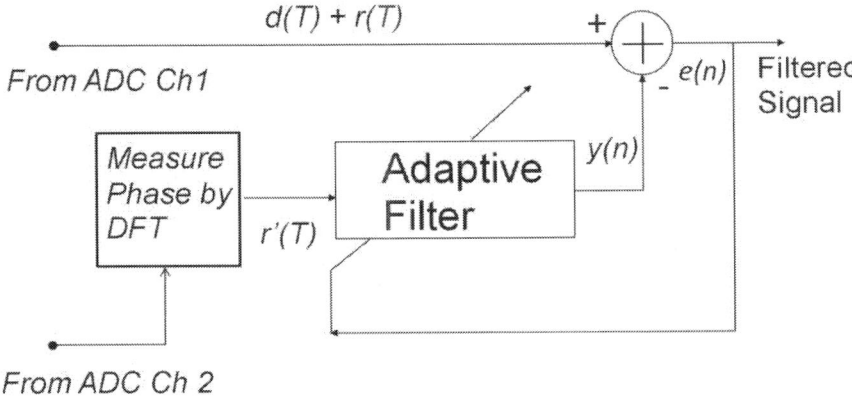

Figure 16. Line beat Adaptive canceller.

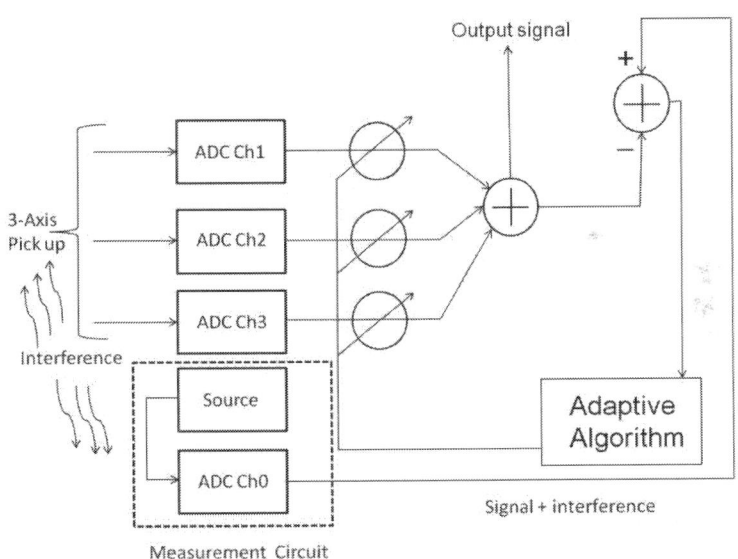

Figure 17. Three Axis linear combiner for interference cancellation

Figure 17 shows a 3-axis magnetic field sensor which is connected to a separate analogue to digital converter (ADC). A forth ADC is used to sample the "signal" simultaneously with the 3-axis data. The three "noise" channel ADCs are the inputs to the three channels of a three-way linear combiner (Wright et al., 2010).

8. CONCLUSION

In recent years, the development and commercial availability of increasingly powerful andaffordable digital computers has been accompanied by the development of advanced digitalsignal processing algorithms for a wide variety of applications; therefore the use of adaptive filters is bigger every day.

Adaptive filters are used forestimation of nonstationary signals and systems, or in applications where a sample-bysampleadaptation of a process and/or a low processing delay is required.

In this chapter, we described some of the most used adaptive filtering applications. The materialpresented here forms the basis to understand the behavior of most adaptive-filtering structuresin practical implementations. The main objective was to illustrate how the adaptive-filtering is applied to solve practical problems.

Thedistinctive feature of each application is the way the adaptive filter input signal and the desiredsignal are chosen. Once these signals are determined, any known properties of them can be usedto understand the expected behavior of the adaptive filter when attempting to minimize the chosenobjective function. The efficiency of the adaptive filters mainly depends onthe used technique of design and the algorithm of adaptation.

REFERENCES

1. M. Bellanger, 2001 Adaptive Digital Filters(Second edition). Marcel Dekker, 0-82470-563-7 York

2. W. . Chen, T. . Nemoto, T. . Kobayashi, T. . Saito, E. Kasuya, Y. Honda, 2000 ECG and Heart RateDetection of Prenatal Cattle Fœtus Using Adaptive Digital Filtering, Engineering in Medicine and Biology Society, 2000. Proceedings of the 22nd Annual International Conference of the IEEE, 2 (July 2000), 123

3. P. Diniz, 2008. Adaptive Filtering, Algorithms and Practical Implementation (Third edition). Springer, ISBN 978-0-387-31274-3, Rio de Janeiro

4. P. . He, G. Wilson, . Rusell, C. , 2004 Removal of ocular artifacts from electro-encephalogram by adaptive filtering, Medical & Biological Engineering & Computing, 42 3 (May 2004), 407412

5. S. . Kuo, B. Lee, W. Tian, 2006 Real Time Digital Signal Processing(Second edition). John Wiley & Sons Ltd, 0-47001-495-4 Sussex

6. G. , F. . Gran, F. Jocobsen, F. Agerkvist, 2010 Adaptive feedback cancellation with band-limited LPC vocoder in digital hearing aids, IEEE Transactions on Audio, Speech, and Language Processing, Vol. pp, 99 (July 2010), 1 1558-7916

7. A. Spriet, G. Rombouts, M. Moonen, J. Wouters, 2006 Adaptive Feedback Cancellation in hearing aids Journal of the Franklin Institute 343 6 (September 2006),545573

8. A. . Shoval, D. Johns, W. Snelgrove, 1995 Comparison of DC Offset Effects in Four LMS Adaptive Algorithms,IEEE Transactions on Circuits and Systems-II: Analog and Digital Signal Processing; 42 3 (March 1995), 176185 .

9. S. Vaseghi, 2006 Advanced Digital Signal Processing and noise reduction (Third edition). John Wiley & Sons Ltd, 047009494West Sussex.

10. P. S. . Wright, P. Clarkson, M. J. Hall, 2010 Application of adaptive noise cancelling filters in ac electrical measurements, 2010 Conference on Precision Electromagnetic Measurements (CPEM), 978-1-42446-795-2 Daejeon, June 2010

Chapter 2

A Robust Weighted Combination Forecasting Method Based on Forecast Model Filtering and Adaptive Variable Weight Determination

Lianhui Li [1,†], Chunyang Mu [2,*,†], Shaohu Ding [1,*], Zheng Wang [3], Runyang Mo [4,5] and Yongfeng Song [4,6]

[1] *College of Mechatronic Engineering, Beifang University of Nationalities, Yinchuan 750021, China*
[2] *State Key Laboratory of Robotics and System, Harbin Institute of Technology (HIT), Harbin 150001, China*
[3] *State Grid Ningxia Electric Power Design Co. Ltd., Yinchuan 750001, China*
[4] *School of Management, Qingdao Technological University, Qingdao 266520, China*
[5] *College of Electrical &Information Engineering, Hunan University, Changsha 410082, China*
[6] *College of Electrical Engineering and Information, Sichuan University, Chengdu 610065, China*

ABSTRACT

Medium-and-long-term load forecasting plays an important role in energy policy implementation and electric department investment decision. Aiming to improve the robustness and accuracy of annual electric load forecasting, a robust weighted combination load forecasting method based on forecast model filtering and adaptive variable weight determination is proposed. Similar years of selection is carried out based on the similarity between the history year and the forecast year. The forecast models are filtered to select the better ones according to their comprehensive validity degrees. To determine the adaptive variable weight of the selected forecast models, the disturbance variable is introduced into Immune Algorithm-Particle Swarm Optimization (IA-PSO) and the adaptive adjustable strategy of particle search speed is established. Based on the forecast model weight determined by improved IA-PSO, the weighted combination forecast of annual electric load is obtained. The given case study illustrates the correctness and feasibility of the proposed method.

Keywords: load forecasting; robustness; combination forecast; Markov chain; normal cloud model; immune algorithm; particle swarm optimization

1. INTRODUCTION

Nowadays, the strong smart grid (SSG) is vigorously being constructed and the renewable distributed electricity generation capacity is steadily increasing. As an important basis to ensure the security and stable operation of the electric system, electric load forecasting is playing a more and more important role in the implementations of energy policies and the investment decision-making of the electric department under this background [1,2,3,4,5,6,7,8,9,10]. However, medium-and-long-term load forecasting has non-linear characteristics caused by the influence of various factors (e.g., national policy, economic and social factors) [7,8,9]. It makes the medium-and-long-term load forecasting much complex and uncertain. Thus, how to improve the robustness and accuracy of annual electric load forecasting is very worthy of study.

On the one hand, the forecast accuracy has reached a high level under the given sample and condition in the existing research [1,2,3,4,5,6,7,8,9,10,11,12,13], but the forecast method's robustness currently becomes the bottleneck. The main reason lies in the status that the existing methods are mostly based on the error theory. By the error theory, the fact of the unknown amount of the forecast year's true load is neglected and the accurate forecast error is difficult to obtain. Chen [14,15] pointed out that the validity degree of the forecast model can be expressed by its full and average precision. In the mathematical sense, any forecast model has its inherent attributes which can be measured by its validity rather than the result error reported by Sun et al. [16], Chen et al. [17] and Jin et al. [18]. At the same time, the single forecast models should be filtered so that the better ones will be selected and the worse ones will be eliminated. Therefore, filtering the forecast models to select the better ones based on their comprehensive validity degrees can improve the robustness of the forecast method.

On the other hand, the general load forecast method mainly includes artificial neural network reported by Hernandez et al. [19] and Gofman et al. [20], regression analysis reported by Li et al. [21] and time series analysis reported by Paparoditis *et al.* [22]. The exponential smoothing method reported by Weron *et al.* [23], the gray forecast method reported by Li *et al.* [24] based on time trend extrapolation reported by Ismail *et al.* [25], the clustering forecast method reported by Kodogiannis *et al.*[26] and multiple regression analysis method reported by Hong *et al.* [27] based on the load related factor analysis cannot ensure the satisfactory result in any case. In order to make full use of the advantages and the contained information of each single forecast model, combination forecast [28,29,30,31,32,33] is an effective method. The question of how to determine the weight assignment of single forecast method is a difficult point in combination forecasting. The constant weight and the variable weight are two common weight determination ways, and the variable weight has a better adaptability. Focusing on this point, a large number of research has been carried out such as mathematical programming method reported by Ma *et al.*[34], genetic algorithm reported by Chaturvedi *et al.* [2], Bayesian method reported by Niu *et al.* [35] and neural network method reported by Hernandez *et al.* [19] and Gofman *et al.* [20]. These methods mostly have the stabile performance and the accuracy meeting the application requirements, but there are still several problems such as much complexity, slow convergence or strong status dependence.

In this paper, we propose a robust weighted combination forecasting method based on forecast model filtering and adaptive variable weight determination. Firstly, the similar years are selected from the sample history years according to the similarity between the history year and the forecast year. Secondly, the forecast models are filtered based on the comprehensive validity degree which is composed of the fitted validity degree in the history year interval and the estimated forecast validity degree in the forecast year interval. Thirdly, the improved Immune Algorithm-Particle Swarm Optimization (IA-PSO), in which the disturbance variable is introduced and the adaptive adjustable strategy of particle search speed is established, is used to determine the adaptive variable weight of the selected forecast models. Lastly, the weighted combination forecast is carried out. The flowchart of the proposed method is shown in Figure 1.

2. SIMILAR YEARS SELECTION

There are a series of factors influencing the annual load. For example, in order to reflect the influence of each factor on the load forecasting result, Zhu *et al.* [36] have investigated an Artificial Neural Network-based approach for medium-and-long-term load forecasting. In the proposed three-layer back propagation network, seven factors are selected as inputs which include Gross Domestic Product (GDP), heavy industry production, light industry production, agriculture production, primary industry, secondary industry and tertiary industry; Wang *et al.* [37] have pointed out that there are mainly eight factors affecting the annual load: area GDP, primary industry GDP ratio, secondary industry GDP ratio, tertiary-industry GDP ratio, power consumption per unit of GDP, electricity price, urban per-capita income and rural per capita income; Lei et al. [38] have analyzed the variation characteristics of the annual maximum load, annual minimum load and typical daily load based on the recent historical load data and meteorological data of Chongqing region. Then, they have studied the interrelationship between load characteristics and major influencing factors. The results show that the temperature, rainfall, holidays and festivals have a significant influence on the region power load; Liao et al. [39] have researched the current load characteristics of Changde region and main factors influencing load variation. Influencing extents of main influencing factors on regional load, respective proportions of these factors in the influences and the time periods influenced by these factors are analyzed, and the quantization analysis on the relation between load and influencing factors is performed. In conclusion, the factors considered by Wang et al. [37] are more than Zhu et al. [36], and the factors considered by Lei et al.[38] and Liao et al. [39] relate to the characteristics in short-term load forecasting. After comparison and summary, we choose the eight factors pointed out by Wang et al. [37] which are more comprehensive and reasonable than others to select the similar years in medium-and-long-term load forecasting.

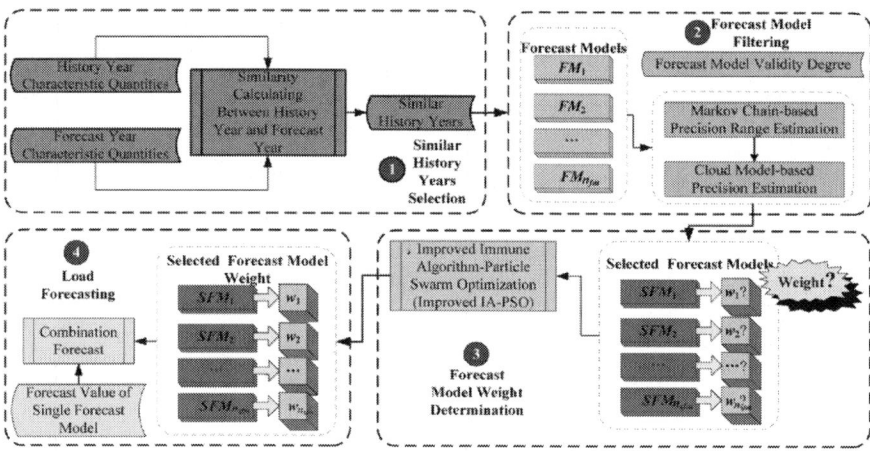

Figure 1. The flowchart of the proposed robust weighted combination forecasting method.

It is assumed that there are n_{his} history years {1, 2, ..., n_{his}} and n_{fo} forecast years {n_{his} + 1, n_{his} + 2, ..., n_{his} + n_{fo}}. The year characteristic is defined as the factor affecting the annual load and the year characteristic quantity is defined as the value of year characteristic in a history year.

If the year characteristic CH_a is efficiency type, the year characteristic quantity $CHQ_{a,b}$ which means the value of the year characteristic CH_a of the history year HY_b is standardized as follows:

$$CHQ_{a,b} = \frac{CHQ_{a,b}}{\max\left\{CHQ_{a,1}, CHQ_{a,2}, \cdots, CHQ_{a,n_{his}}\right\}} \tag{1}$$

where $a = 1, 2, ..., n_{ch}$, $b = 1, 2, ..., n_{his}$.

If the year characteristic CH_a is cost type, the year characteristic quantity $CHQ_{a,b}$ which means the value of the year characteristic CH_a of the history year HY_b is standardized as follows:

$$CHQ_{a,b} = \frac{\min\left\{CHQ_{a,1}, CHQ_{a,2}, \cdots, CHQ_{a,n_{his}}\right\}}{CHQ_{a,b}} \tag{2}$$

where $a = 1, 2, ..., n_{ch}$, $b = 1, 2, ..., n_{his}$.

For two individuals whose characteristics have the same dimension number, distance and similarity are usually used to measure the difference between them. The distance measures the absolute distance between two individuals in space which is directly related to the position coordinates of each individual (i.e., the numerical value of each characteristic dimension), but the similarity measures the angle between two individual vectors and reflects more difference in direction that in difference in position [40,41,42]. Therefore, similarity is more suitable to measure the difference between a history year and a forecast year.

Here, the most common cosine similarity is chosen [41,42]. The cosine similarity between the history year HY_b and the forecast year FY_c is as follows:

$$CSI_{b,c} = \frac{\sum\limits_{a=1}^{n_{ch}} CHQ_{a,b} \cdot CHQ_{a,c}}{\sqrt{\left(\sum\limits_{a=1}^{n_{ch}} CHQ_{a,b}^2\right) \cdot \left(\sum\limits_{a=1}^{n_{ch}} CHQ_{a,c}^2\right)}} \tag{3}$$

where $b = 1, 2, ..., n_{his}$, $c = n_{his} + 1, n_{his} + 2, ..., n_{his} + n_{fo}$.

Lastly, n_{shis} ($n_{shis} < n_{his}$) history years with the highest similarity are selected as the similar years of the forecast year FY_c.

3. FORECAST MODEL FILTERING

3.1. Forecast Model Validity Degree

As shown in Figure 2, the load sequence of similar years is {y'1,y'2,···,y'ns his}, and the load sequence of forecast years is {ynhis+1,ynhis+2,...,ynhis+nfo}. We assume that there are n_{fm} forecast models. By the forecast model FM_d (d = 1, 2, ..., n_{fm}), y'e,d is the fitted value of the similar year e (e = 1, 2, ..., n_{shis}), and yc,d is the forecast value of the forecast year c ($c = n_{his} + 1, n_{his} + 2, ..., n_{his} + n_{fo}$).

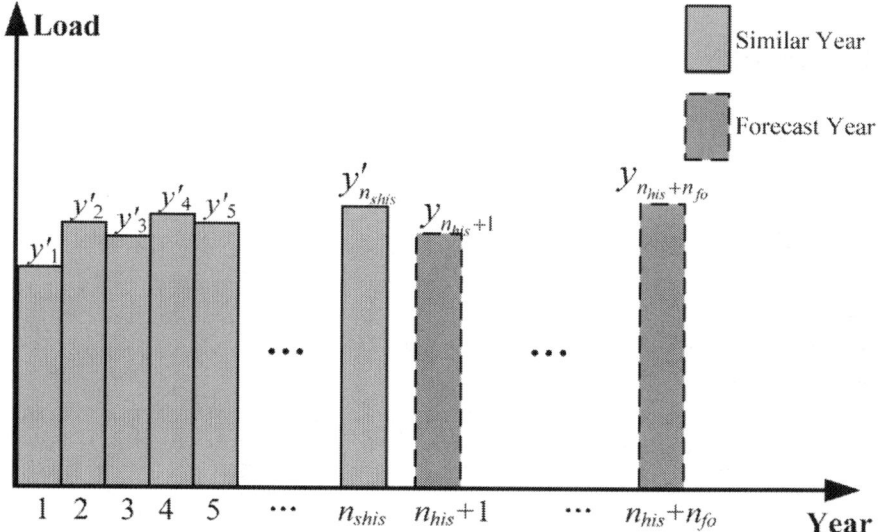

Figure 2. The history years and the forecast years.

The fitted value relative error of FM_d for the similar year e is as follows:

$$RE'_{e,d} = \frac{y'_e - y'_{e,d}}{y'_e} \tag{4}$$

The forecast value relative error of FM_d for the forecast year c is as follows:

$$RE_{c,d} = \frac{y_c - y_{c,d}}{y_c} \tag{5}$$

Then, the fitted precision of FM_d for the similar year e is as follows:

$$P'_{e,d} = \begin{cases} 1-\left|RE'_{e,d}\right|, & 0 \leqslant \left|RE'_{e,d}\right| \leqslant 1 \\ 0, & \left|RE'_{e,d}\right| \geqslant 1 \end{cases} \tag{6}$$

The forecast precision of FM_d for the forecast year c is as follows:

$$P_{c,d} = \begin{cases} 1-\left|RE_{c,d}\right|, & 0 \leqslant \left|RE_{c,d}\right| \leqslant 1 \\ 0, & \left|RE_{c,d}\right| \geqslant 1 \end{cases} \tag{7}$$

Lastly, the fitted validity degree of FM_d is as follows:

$$FIV_d = \mathrm{EXP}(P'_{e,d}) \cdot (1 - \sigma(P'_{e,d})) \tag{8}$$

where $\mathrm{EXP}(P'_{e,d}) = \dfrac{1}{n_{shis}} \displaystyle\sum_{e=1}^{n_{shis}} P'_{e,d}$ and $\sigma(P'_{e,d}) = \sqrt{\dfrac{1}{n_{shis}} \displaystyle\sum_{e=1}^{n_{shis}} (P'_{e,d} - \mathrm{EXP}(P'_{e,d}))^2}$ a re the expectation and the standard deviation of the fitted precision of FM_d for the similar year e, respectively.

The forecast validity degree of FM_d is as follows:

$$FOV_d = \mathrm{EXP}(P_{c,d}) \cdot (1 - \sigma(P_{c,d})) \tag{9}$$

where $\mathrm{EXP}(P_{c,d}) = \dfrac{1}{n_{fo}} \displaystyle\sum_{c=n_{his}+1}^{n_{his}+n_{fo}} P_{c,d}$, $\sigma(P_{c,d}) = \sqrt{\dfrac{1}{n_{fo}} \displaystyle\sum_{c=n_{his}+1}^{n_{his}+n_{fo}} (P_{c,d} - \mathrm{EXP}(P_{c,d}))^2}$ are the expectation and the standard deviation of the forecast precision of FM_d for the forecast year c, respectively.

3.2. Forecast Model Precision Estimation

The validity degree of a forecast model is defined in Equation (9) which can depict its credibility and is a reflection of its inherent property [16,17,18]. Obviously, true value has not yet occurred and the forecast error cannot be obtained in the future forecast interval. We can only estimate the precision and the validity of a forecast model based on its inherent property. Then, the suitable forecast models are selected and the combination forecast model is put forward.

3.2.1. Markov Chain-Based Precision Range Estimation

As an inherent property, the forecast model precision is shown in the form of the fitted precision which is obtained through the virtual forecast for the multi-time load. Using the forecast model FM_d to forecast the load of the similar year e ($e=$ 1, 2, ..., n_{shis}), we can obtain the fitted precision sequence $\left\{P'_{1,d}, P'_{2,d}, \cdots, P'_{n_{shis},d}\right\}$. In this sequence, the expectation and the standard deviation of the fitted precision of each similar year show the property of the forecast model. As is known, randomness and discreteness appear in the fitted precision sequence. Therefore, we can use the Markov chain transition matrix [43,44] to represent the transition rule as follows:

1. The fitted precision distribution interval of FM_d for the similar year e is divided into n_{si} ($n_{si} \leq n_{shis}$) sub-intervals with equal distance as $S_d^1, S_d^2, ..., S_d^{n_{si}}$, where

$$S_d^g = \left[\underline{S_d^g}, \overline{S_d^g}\right], \underline{S_d^1} = \min\left\{P'_{1,d}, P'_{2,d}, \cdots, P'_{n_{shis},d}\right\}, \overline{S_d^{n_{si}}} = \max\left\{P'_{1,d}, P'_{2,d}, \cdots, P'_{n_{shis},d}\right\},$$
$$\overline{S_d^g} - \underline{S_d^g} = \overline{S_d^h} - \underline{S_d^h} \quad s.t.\ g \neq h,\ g, h = 1, 2, ..., n_{si} \tag{10}$$

   Each fitted precision sub-interval can be considered as a fitted precision state.

2. According to the fitted precision of FM_d for the similar year e, the occurrence number of the fitted precision state Sgd is OC^g ($OC^g < n_{shis}$). That means there are OC^g times which belong to the fitted precision state Sdg. The transition number from the fitted precision state S_d^g is OC^g ($OC^g < n_{shis}$). Thus, the transition probability of FM_d from the fitted precision state S_d^h to S_d^g can be obtained as follows:

$$TP_d^{h,g} = \left\{S_d^g = c_g \middle| S_d^h = c_k\right\} = \frac{TRN^{h,g}}{OC^g} \tag{11}$$

According to Equation (11), the one-step state transition matrix of FM_d is as follows:

$$TM_d^{(1)} = \begin{bmatrix} TP_d^{1,1} & TP_d^{1,2} & \cdots & TP_d^{1,n_{si}} \\ TP_d^{2,1} & TP_d^{2,2} & \cdots & TP_d^{2,n_{si}} \\ \vdots & \vdots & & \vdots \\ TP_d^{n_{si},1} & TP_d^{n_{si},2} & \cdots & TP_d^{n_{si},n_{si}} \end{bmatrix} \tag{12}$$

The q-step state transition matrix is as follows:

$$TM_d^{(q)} = (TM_d^{(1)})^q \qquad (13)$$

(3) We construct an initial vector IV_d whose elements are the occurrence numbers of each fitted precision state of FM_d. Though multiplying the initial vector IV_d with the q-step state transition matrix TM(q)d, the new state matrix of FM_d is obtained as follows:

$$SM_d = IV_d \cdot TM_d^{(q)} \qquad (14)$$

(4) We calculate the sum of each column vector of SM_d. If the column vector CVE_i (i = 1, 2, ..., n_{si}) has the maximum sum, the forecast precision will belong to the precision state Sid.

3.2.2. Cloud Model-Based Precision Estimation

Due to the various factors, the fitted precision sequence $P'_d = \left\{ P'_{1,d}, P'_{2,d}, ..., P'_{n_{shis},d} \right\}$ of the forecast model FM_d in n_{shis} similar history years clearly has the property of random variables. As a result, the forecast precision is uncertain in its precision range. Therefore, we can describe this uncertainty by the expectation, entropy and hyper-entropy in the precision range and make the quantitative precision estimation based on the normal cloud model [45,46] as follows:

(1) Firstly, we construct a backward cloud generator. We can map the fitted precision sequence $P'_d = \left\{ P'_{1,d}, P'_{2,d}, ..., P'_{n_{shis},d} \right\}$ into the normal cloud model. In this normal cloud model, the input is P'_d and the output is the expectation Ex_d, the entropy En_d and the hyper-entropy He_d. The algorithm of this backward cloud generator is as follows:

$$Ex_d = EXP\left(P'_d\right) \qquad (15)$$

$$En_d = \sqrt{\frac{\pi}{2}} \cdot \frac{1}{n_{shis}} \sum_{e=1}^{n_{shis}} \left| P'_{e,d} - Ex_d \right| \qquad (16)$$

$$He_d = \sqrt{\frac{1}{n_{shis}-1} \sum_{e=1}^{n_{shis}} (P'_{e,d} - Ex_d)^2 - (En_d)^2} \qquad (17)$$

(2) Secondly, we construct a forward cloud generator. The input is Ex_d, En_d, He_d and the constraint is the precision range S_g^d. The algorithm using the forward cloud generator for precision estimation is as follows:

$$P_{c,d} = \text{NORM}(Ex_d, En'_d) \tag{18}$$

where $En'_d = \text{NORM}(En_d, He_d)$ is a normal random number with the expectation En_d and the variance He_d, $P_{c,d}$ is the estimated forecast precision with the expectation Ex_d and the variance En'_d in the precision range S^d_g.

3.3. Forecast Model Filtering Based on Comprehensive Validity Degree

The comprehensive validity degree of the forecast model FM_d in the whole interval $[1, n_{shis}] \cup [n_{his} + 1, n_{his} + n_{fo}]$ is as follows:

$$CV_d = \alpha \cdot FIV_d + (1 - \alpha) \cdot FOV_d \tag{19}$$

where FIV_d is the fitted validity degree of FM_d in the similar history years interval $[1, n_{shis}]$, FOV_d is the estimated forecast validity degree of FM_d in the forecast years interval $[n_{his} + 1, n_{his} + n_{fo}]$.

The empirical coefficient $\alpha \ (0 \leqslant \alpha \leqslant 1)$ is determined by the forecast staff based on their experiences. The bigger α is, the more important FIV_d is. Here $\alpha = 0.5$, which means that FIV_d and FOV_d are equally important.

We use the mean comprehensive validity degree of n_{fm} forecast models as the forecast model filtering threshold:

$$\overline{CV} = \sum_{d=1}^{n_{fm}} CV_d \tag{20}$$

If $CV_d \geqslant \overline{CV}$, the forecast model FM_d will be selected for the combination forecast, else it will be eliminated. In the applications, the filtering threshold can be adjusted according to the actual situation and forecast decision-makers' experiences.

4. FORECAST MODEL WEIGHT DETERMINATION AND COMBINATION FORECAST

4.1. Mathematical Description

After the forecast models filtering, n_{sfm} ($n_{sfm} < n_{fm}$) better forecast models are selected from n_{fm} forecast models for combination forecast.

We assume that the weight of the selected forecast model SFM_j ($j = 1, 2, ..., n_{sfm}$) is ωj ($\sum_{j=1}^{n_{sfm}} \omega j = 1$, $0 \leq \omega j \leq 1$). For the similar history year e ($e = 1$,

2, ..., n_{shis}), the actual load is y'e and the forecast load by SFM_j is $y'''_{e,j}$ so the combination forecast load of n_{sfm} selected forecast models is as follows:

$$y'''_e = \sum_{j=1}^{n_{sfm}} \omega_j y'''_{e,j} \tag{21}$$

The target is to achieve the minimum square sum of the combination forecast error, so the mathematical description is described as follows:

$$\min \sum_{e=1}^{n_{shis}} \left(\sum_{j=1}^{n_{sfm}} \omega_j y'''_{e,j} - y'_e \right)^2, \quad \text{s.t.} \sum_{j=1}^{n_{sfm}} \omega_j = 1, 0 \leqslant \omega_j \leqslant 1 \tag{22}$$

4.2. Improved Immune Algorithm-Particle Swarm Optimization (Improved IA-PSO)

4.2.1. Particle Swarm Optimization (PSO)

As a kind of stochastic optimization algorithm, particle swarm optimization (PSO) [47] is developed based on the simulation of bird-group foraging behavior. To search for the optimal solution, the individuals have to cooperate and compete with each other in PSO.

There is an m-dimensional space and an initial swarm PO composed of n_{pa} particles as follows:

$$PO = \left\{ PA_1, PA_2, \ldots, PA_{n_{pa}} \right\} \tag{23}$$

For the particle PA_l ($l = 1, 2, \ldots, n_{pa}$), its position x_l and its speed v_l are expressed by two m-dimensional vectors as follows:

$$x_l = \left(x_{l,1}, x_{l,1}, \ldots, x_{l,m} \right)^T \tag{24}$$

$$v_l = \left(v_{l,1}, v_{l,1}, \ldots, v_{l,m} \right)^T \tag{25}$$

Each particle is moving in the solution space, and its direction is determined by its speed. The speed and position of the particle is continuously updated as follows:

$$v_l^{(k+1)} = w \cdot v_l^{(k)} + LF_1 \cdot rand_1^{(k)} \cdot (pbest_l^{(k)} - x_l^{(k)}) + LF_2 \cdot rand_2^{(k)} \cdot (Gbest_l^{(k)} - x_l^{(k)}) \tag{26}$$

$$x_l^{(k+1)} = x_l^{(k)} + v_l^{(k+1)} \tag{27}$$

where $v_I^{(k)}$ and $x_I^{(k)}$ are the speed and position of the particle PA_I in the iteration $iter_k$, LF_1 and LF_2 are the learning factors, and $0 \le rand_1^{(k)}, rand_2^{(k)} \le 1$ are two random numbers.

- The momentum part $w \cdot v_I^{(k)}$ represents the trust in its current motion state where w is the inertia coefficient used to control the influence of the speed $v_I^{(k)}$ on the speed $v_I^{(k+1)}$. This part provides a necessary momentum which enables the particle to carry on the inertia motion based on its speed.

- The individual cognitive part $LF_1 \cdot rand_1^{(k)} \cdot (pbest_I^{(k)} - x_I^{(k)})$ represents the particle self-thinking behavior. This part encourages the particle to fly to the best position found by itself.

- The social cognitive part $LF_2 \cdot rand_2^{(k)} \cdot (Gbest_I^{(k)} - x_I^{(k)})$ represents the information sharing and cooperation of different particles. This part guides the particle to fly to the best position of the group.

Therefore, the momentum part $w \cdot v_I^{(k)}$ represents the particles' diversification; the individual cognitive part $LF_1 \cdot rand_1^{(k)} \cdot (pbest_I^{(k)} - x_I^{(k)})$ and the social cognitive part $LF_2 \cdot rand_2^{(k)} \cdot (Gbest_I^{(k)} - x_I^{(k)})$ represent the particles' centralization. The main performance of PSO is determined by the balance of the three parts.

In early evolution, PSO has the advantage of fast convergence speed and simple operation, so it can be used for solving the nonlinear, non-differentiable and multi-peak complex optimization problems. But in late evolution, PSO has a significantly slower convergence speed and reaches a poor accuracy, so it is easy to fall into the local optimum.

4.2.2. Immune Algorithm-Particle Swarm Optimization (IA-PSO)
To solve this problem, IA-PSO introduces biological immune system's specific information processing mechanism (e.g., immune memory, immune regulation and immune selection) into PSO's basic framework [48,49].
- Immune memory: The immune system keeps the antibodies opposing against the invading antigen as memory cells. If the same antigen invades again, the memory cells will be activated and produce a large number of antibodies. In IA-PSO, this idea is used to preserve the excellent particle. The best position $pbest_I^{(k)}$ searched by each particle up to now is considered as a memory cell. If the new born particles are detected not to meet the requirements, they will be replaced by the memory cells.

- Immune regulation: In IA-PSO, immune regulation is used for particle selection. If a particle has a strong affinity or a low concentration, it will be promoted. Otherwise, it will be demoted. Therefore, the particle diversification can be always kept. The selected probability [48,49] of the particle PA_I is as follows:

$$PRO_I = \chi \cdot PRO_{I1} + (1 - \chi) \cdot PRO_{I2} \qquad (28)$$

In Equation (28), $PRO_{I1} = AF_I / \sum_{u=1}^{all} AF_u$ represents the selected probability determined by the affinity where AF_I is the affinity of the particle PA_I, $PRO_{I2} = CON_I^{-1} / \sum_{u=1}^{all} CON_u^{-1}$ represents the selected probability determined by the concentration where CON_I is the concentration of the particle PA_I. χ represents the weight of PRO_{I1} and $1 - \chi$ represents the weight of PRO_{I2} ($0 \leq \chi \leq 1$).

- Immune selection: In the immune system, vaccinating means to change several components of the antibody according to the vaccination. In IA-PSO, the group best position $Gbest_i^{(k)}$ up to the iteration $iter_k$ can be considered as the closest one to the optimal solution. Thus, we use several components of $Gbest_i^{(k)}$ as the vaccination to vaccinate the particles and calculate the particle fitness value for immune selection. If the particle fitness value after the vaccination is lower than its parent, the vaccination will be abolished. Otherwise, the particle will be retained.

IA-PSO, which inherits the global optimization ability of PSO and the immune information processing mechanism of IA, can improve the algorithm accuracy. But at the same time, the algorithm complexity is increased because of the introduction of the immune system.

4.2.3. Improved IA-PSO Based on Disturbance Variable

By introducing a disturbance variable and establishing the searching speed adaptive mechanism, we improve IA-PSO in this paper. Through this improvement, not only the diversity of particles can be ensured to avoid the local optimum, but also the accuracy and convergence speed can be increased.

To solve the mathematical problem described in Equation (22), the search space is set as n_{sfm}-dimensional, the particles number is n_{pa}, and the maximum iteration number is $iter_{Max}$. The position of each particle is an n_{sfm}-dimensional vector in which each dimensional represents the weight of a selected forecast model. After the iteration $iter_k$ ($iter_k = 1, 2, ..., iter_{Max}$), the position and speed of the particle PA_l ($l = 1, 2, ..., n_{pa}$) are as follows:

$$x_l^{(k)} = (x_{l,1}^{(k)}, x_{l,2}^{(k)}, \ldots, x_{l,n_{sfm}}^{(k)})^T \tag{29}$$

$$v_l^{(k)} = (v_{l,1}^{(k)}, v_{l,2}^{(k)}, \ldots, v_{l,n_{sfm}}^{(k)})^T \tag{30}$$

$pbest_l^{(k)}$, $gbest^{(k)}$, $Gbest^{(k)}$ are used to represent the best position searched by the particle PA_l in the iteration $iter_k$, the best position searched by the particle swarm in the iteration $iter_k$, the best position searched by the particle swarm up to the iteration $iter_k$, respectively.

Introducing Disturbance Variable into IA-PSO

In the production process of a new particle swarm, the particle position is updated according to Equation (27), and the step length is represented by Equation (26). The coefficients of the three parts in Equation (26) are randomly changed, but only the rules of the particle to follow the order are changed. It means that the step length only reflects the rules of the search behavior and the diversification is lacking. Group optimization should be a balance between the order following and random irrational behaving, so we introduce the disturbance term [50,51] to Equation (26). In each iteration, the particle position is updated as follows:

$$x_l^{(k)} = x_l^{(k-1)} + v_l^{(k)} + (rand_3^{(k)} - 0.5) \cdot \beta_l^{(k)} \tag{31}$$

where $0 \leqslant rand_3^{(k)} \leqslant 1$ is a random number subject to uniform distribution, $\beta_l^{(k)} > 0$ is the disturbance variable of the particle PA_l in the iteration $iter_k$. The disturbance term $(rand_3^{(k)} - 0.5) \cdot \beta_l^{(k)}$ reflects an unpredictable random behavior.

The disturbance variable $\beta_l^{(k)}$ controls the random decision-making behavior strength of the particle PA_l in the iteration$iter_k$. If $\beta_l^{(k)}$ is too big, the awareness of order following will be submerged. If $\beta_l^{(k)}$ is too small, the population diversity and global search ability will be reduced. Therefore, the disturbance variable of each particle should be adjusted in the algorithm operation according to its evolution speed. The adjustment can make the particle swarm with a good generalization ability and convergence speed in the evolution process. Thus $\beta_l^{(k)}$ is defined as follows:

$$\beta_l^{(k)} = \beta_{min} \cdot \frac{F\left(pbest_l^{(k-1-\theta)}\right) - F\left(pbest_l^{(k-1-\rho)}\right)}{F\left(pbest_l^{(k-1)}\right) - F\left(pbest_l^{(k-1-\theta)}\right)}$$

$$\text{s.t.} \quad pbest_l^{(k-1)} \neq pbest_l^{(k-1-\theta)} \neq pbest_l^{(k-1-\rho)}$$

$$\theta < \rho, \theta = \min\{1,2,\ldots,k\}, \rho = \min\{1,2,\ldots,k\}$$

(32)

where β_{min} is the minimum value of the disturbance variable, $F(\cdot)$ is the fitness function, $pbest_l^{(k-1)}$, $pbest_l^{(k-1-\theta)}$, $pbest_l^{(k-1-\rho)}$ is the best position found by the particle PA_l in the iteration $iter_k-1$, $iter_k-1-\theta$, $iter_k-1-\rho$. When the evolution begins, β_l is bigger which means that the particle PA_l has a step length with strong randomness. After several iterations, β_l tend to β_{min} which means that the step length randomness of the particle PA_l becomes weak.

Through the introduction of the disturbance term, Equation (31) reflects the positive and negative sides of the particle updating decision. In Equation (31), the first part $x_l^{(k-1)}$ is the original position, the second part $v_l^{(k)}$ reflects the step length of order following, and the third part $(rand_3^{(k)} - 0.5) \cdot \beta_l^{(k)}$ reflects the step length of random irrational behaving. Due to the disturbance variable, the particle position updating can be ensured and a strong search desire can be kept even if the local optimum appears when compared with Equation (26). As a result, the premature convergence problem can be overcome, the local best solution can be prevented and the algorithm accuracy can be improved.

Establishing the Adaptive Adjustable Strategy of Particle Searching Speed

In the particle searching process, the searching speed should be adaptively adjusted to accelerate the convergence based on the diversity of particles. For the excellent particles, their searching speeds should be decreased to make them quickly be close to the global best solution, and then the convergence can be accelerated. For the poor particles, their searching speed should be adjusted according to the convergence degree of the particle swarm: if the swarm individuals tend to be dispersed, the searching speed should be reduced and the swarm development ability should be enhanced to strengthen the local optimization; if the swarm individuals tend to be converged (the algorithm falls into local optimum), the searching speed should be increased and the swarm detection ability should be enhanced to effectively jump out of the local optimum and achieve the accelerated convergence [42,43].

In iteration $iter_k$, the fitness of the particle PA_l is $PF_l^{(k)}$, the fitness of the best particle is $PF_0^{(k)}$, the average fitness of the swarm is $PF_{avg}^{(k)}$, and the average fitness of the particles whose fitness are bigger than $PF_{avg}^{(k)}$ is $PF_{AVG}^{(k)}$. We use $\Delta^{(k)} = \left| PF_0^{(l)} - PF_{AVG}^{(h)} \right|$ to evaluate the swarm convergence degree in the iteration L. According to the particle fitness, the swarm is divided into three sub-swarms: $PF_l^{(k)} > PF_{AVG}^{(k)}$, $PF_{avg}^{(k)} < PF_l^{(k)} < PF_{AVG}^{(k)}$, $PF_l^{(k)} < PF_{avg}^{(k)}$. For the different sub-swarms, we take different adjustment operations to their searching speeds as follows:

$$(1)\ PF_l^{(k)} > PF_{AVG}^{(k)}$$

As the excellent individuals in the swarm, these particles have been relatively close to the global optimum. Their searching speeds should be reduced to prevent from jumping out of the global optimum. So the searching speed is adjusted as follows:

$$v_{new,l}^{(k)} = \left(1 - 0.5\frac{PF_l^{(k)} - PF_{AVG}^{(k)}}{\left|\Delta^{(k)}\right|}\right) \cdot v_l^{(k)} \tag{33}$$

If the particle is more excellent, it will have a lower searching speed. As a result, the local optimum ability is strengthened and the convergence is accelerated.

$$(2)\qquad PF_{avg}^{(k)} < PF_l^{(k)} < PF_{AVG}^{(k)}$$

As the general individuals of the swarm, both the local optimum ability and the global optimum ability of these particles are good. Therefore, we don't adjust their searching speeds.

$$(3)\qquad PF_l^{(k)} < PF_{avg}^{(k)}$$

These particles are the relatively poor individuals in the swarm. The searching speed is adjusted as follows:

$$v_{new,l}^{(k)} = \left(1.5 - \frac{1}{1 + \eta_1 \cdot \exp(-\eta_2 \cdot \Delta^{(k)})}\right) \cdot v_l^{(k)} \tag{34}$$

where $\eta_1, \eta_2 > \eta_1, \eta_2 > 0$ and η_1 is used to control the upper limit of $v_{new,l}^{(k)}$. If η_1 is bigger, the upper limit of $v_{new,l}^{(k)}$ will be bigger. Here we choose $\eta_1 = \eta_2 = 4$. $\Delta^{(k)} \geq 0$, so $v_{new,l}^{(k)} \in \left[0.5 \cdot v_i^{(L)}, 1.3 \cdot v_i^{(L)}\right]$.

In the searching process, $v_{new,l}^{(k)}$ of these particles is dynamically and adaptively adjusted according to the value of $\Delta^{(k)}$: if the individuals tend to be dispersed ($\Delta^{(k)}$ becomes bigger), $v_{new,l}^{(k)}$ will be reduced and the swarm development ability will be enhanced to strengthen the local optimization; if the individuals tend to be converged ($\Delta^{(k)}$ becomes smaller), $v_{new,l}^{(k)}$ will be increased and the swarm detection ability will be enhanced to effectively jump out of the local optimum.

4.3. Implementation Steps of Forecast Model Weight Determination Based on Improved IA-PSO

The flowchart of forecast model weight determination (FMWD) based on improved IA-PSO is shown in Figure 3.

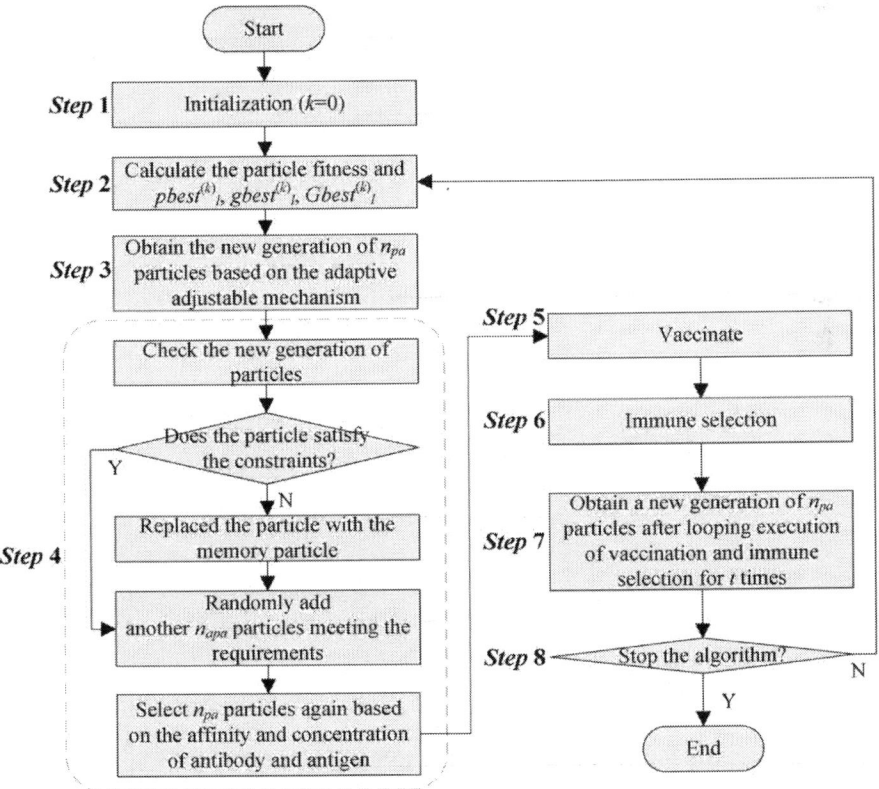

Figure 3. The flowchart of forecast model weight determination based on improved IA-PSO.

The implementation steps are as follows:

Step 1: Initialization. It is assumed that the elements of the particle position vector x all belong to the interval [0, 1], the elements of the particle speed vector v all belong to the interval $[-v_{max}, v_{max}]$, the maximum iteration number is $iter_{max}$ and the initial value of the iteration number k is k = 0. n_{pa} particles are randomly generated. The particle PA_l (l = 1, 2, ..., n_{pa}) has the position x(0)l and the flying speed $v_l^{(0)}$.

Step 2: Calculate $pbest_l^{(k)}, gbest_l^{(k)}, Gbest_l^{(k)}$ and the fitness of the particle PA_l. Here, the fitness of the particle PA_l can be represented by the target function in Equation (22) as follows:

$$F_l^{(k)} = \sum_{e=1}^{n_{shis}} \left(\sum_{j=1}^{n_{sfm}} x_{l,j}^{(k)} \cdot y_{e,j}'' - y_e' \right)^2 \tag{35}$$

Step 3: Obtain the new generation of particles. The particle speed is adjusted based on the adaptive adjustable mechanism and the new position $x_l^{(k+1)}$ and new speed $v_l^{(k+1)}$ can be obtained according to Equations (26) and (31). The elements of the new speed $v_l^{(k+1)}$ must belong to the interval $[-v_{max}, v_{max}]$.

Step 4: Check the new generation of particles. If the position of a new particle is an infeasible solution (one or more elements of the new position $x_l^{(k+1)}$ don't belong to the interval [0, 1]), this particle will be replaced with the memory particle $pbest_l^{(k)}$.

In addition, another n_{apa} particles meeting the requirements will be randomly added. According to Equation (28), n_{pa} particles are selected from the $n_{pa} + n_{apa}$ particles based on the affinity and concentration of antibody and antigen.

Step 5: Vaccinate. One particle is randomly selected from n_{pa} new particles. Then, one element is randomly selected from the front $t - 1$ elements of $Gbest_l^{(k)}$ to exchange with the selected particle at the corresponding element. The tth element of the selected particle is calculated as follows:

$$x_{l,t}^{(k+1)} = 1 - \sum_{j=1}^{t-1} x_{l,j}^{(k+1)} \tag{36}$$

Hence the vaccination is finished.

Step 6: Immune selection. The particle fitness after the vaccination is calculated. If the particle fitness value after the vaccination is lower than its parent, the vaccination will be abolished. Else, the particle will be retained.

Step 7: Looping execution of Steps 4 and 5 for r times (r times vaccination). A new generation of n_{pa} particles is obtained.

Step 8: Judge whether the algorithm should be stopped. The stopping of algorithm is usually determined by the maximum iteration number or the precision. If the algorithm meets its stopping condition, the optimization will be stopped. Else, $k = k + 1$ and go back to Step 2 to continue.

4.4. Weighted Combination Forecast

Based on the improved IA-PSO, the weight ω_j of the selected forecast model SFM_j ($j = 1, 2, ..., n_{sfm}$) is obtained where $\sum_{j=1}^{n_{sfm}} \omega_j = 1, 0 \leqslant \omega_j \leqslant 1.$ For the selected forecast model SFM_j, $y''_{c,j}$ is the forecast value of the forecast year c ($c = n_{his} + 1, n_{his} + 2, ..., n_{his} + n_{fo}$). So the weighted combination forecast (WCF) value of the forecast year c is as follows:

$$y''_c = \sum_{j=1}^{n_{sfm}} \omega_j \cdot y''_{c,j} \tag{37}$$

5. CASE STUDY

We use the proposed method to forecast a Chinese province's load of 2005 based on its loads of the history years from 1998 to 2004. The year characteristic quantity data from 1998 to 2005 is shown in Table 1.

The cosine similarity between the history years 1998–2004 and the forecast year 2005 is shown in Table 2.

Table 1. The year characteristic quantity data from 1998 to 2005 [37,52].

Year	Area GDP (10^8 Yuan)	Primary Industry GDP Ratio (%)	Secondary Industry GDP Ratio (%)	Tertiary Industry GDP Ratio (%)	Power Consumption per Unit of GDP (kWh/Yuan)	Electricity Price (Yuan/kWh)	Urban per-Capita Income (Yuan)	Rural per-Capita Income (Yuan)
1998	9686.6	12.95	50.22	36.83	0.156	0.408	3005.21	1896.56
1999	9802.8	12.90	49.90	37.20	0.152	0.408	3859.86	2003.63
2000	9912.3	12.88	50.07	37.05	0.150	0.410	4663.23	2150.36
2001	10,626.6	12.83	50.06	37.11	0.148	0.412	5551.91	2340.14
2002	11,386.5	12.80	49.69	37.51	0.142	0.412	6599.24	2485.86
2003	12,955.2	12.38	50.75	36.87	0.139	0.412	7370.65	2657.93
2004	15,133.9	12.68	51.63	35.69	0.132	0.419	8245.55	3103.98
2005	17,140.8	12.79	49.62	37.59	0.125	0.444	9227.55	3391.82

Table 2. The cosine similarity between the history years 1998–2004 and the forecast year 2005.

Year	Cosine Similarity (%)
1998	98.56
1999	95.21
2000	94.66
2001	99.60
2002	97.52
2003	96.37
2004	98.63

According to forecasting decision-makers' experiences, we choose the threshold value of similar years selection as 96%. Therefore, the five history years (1998 and 2001–2004) with highest similarity are selected as the similar years of the forecast year 2005.

The power consumption of the province in 1998 and 2001–2005 is shown in Table 3.

Table 3. The power consumption of the Chinese province in 1998 and 2001–2005 (10^8kWh) [16,52].

Year	1998	2001	2002	2003	2004	2005
Power Consumption	437.85	557.58	628.82	725.20	833.01	946.33

The forecast values by eleven forecast models are shown in Table 4.

By the validity degree calculation method based on Markov chain and cloud model, the comprehensive validity degrees of the eleven forecast models are obtained as shown in Table 5.

The forecast model filtering threshold of the eleven forecast models is $\overline{cvd} = 81.46\%$. Therefore, the forecast models FM_4, FM_5, FM_8, FM_9, FM_{11} are selected for the combination forecast and the others are eliminated. Respectively, we use PSO, IA-PSO and improved IA-PSO for the forecast model weight determination. There are five selected forecast models SFM_1 (FM_4: Power function model), SFM_2 (FM_5: S-curve model), SFM_3 (FM_8: Cubic curve model), SFM_4 (FM_9: Artificial neural network method) and SFM_5 (FM_{11}: Grey system method). The parameters are set as follows: n_{pa} = 100, n_{sfm} = 5, v_{max}= 1, $iter_{max}$ = 1000, w = 0.6, $LF_1 = LF_2$ = 2, n_{apa} = 30, r = 25, βmin = 0.001, η1 = η2 = 4. The results of forecast model weights determination (FMWD) by PSO, IA-PSO and improved IA-PSO, which are abbreviated as FMWD-PSO, FMWD-IA-PSO, FMWD-improved-IA-PSO, are shown in Table 6.

Table 4. The forecast power consumption by the eleven forecast models (10^8 kWh).

Forecast Model	1998	2001	2002	2003	2004	2005
FM_1: Exponential model ($y = 780.65e^{-0.82/x}$)	343.91	636.02	662.58	680.93	694.36	704.55
FM_2: Logarithm model ($y = 362.13 + 188.39\ln x$)	362.09	623.33	665.32	699.65	728.73	753.90
FM_3: Hyperbola model ($y = 722.84 - 354.37/x$)	368.54	634.24	652.01	663.76	672.21	678.53
FM_4: Para-curve model ($y = 431.79 - 3.58x + 8.17x^2$)	436.88	556.81	631.53	723.67	833.29	960.30
FM_5:Grey system method [53]	437.94	567.04	642.01	727.00	823.23	932.08
FM_6: COMPERTZ model ($\ln y = 6.46 - 1.29e^{-x}$)	400.04	626.88	636.27	639.75	641.14	641.64
FM_7: Power function model ($y = 358.90x^{0.385}$)	358.90	612.43	667.48	716.12	759.91	800.02
FM_8: Cubic curve model ($y = 432.10 - 3.94x + 8.81x^2 - 0.0087x^3$)	437.04	556.81	631.56	723.84	833.17	960.04
FM_9: Artificial neural network method [36]	406.81	583.02	658.81	725.38	775.44	809.03
FM_{10}: S-curve model ($y^{-1} = 0.0015 + 0.0039e^{-x}$)	345.60	649.47	669.01	676.35	679.22	680.30
FM_{11}: Exponential smoothing method [54]	437.85	544.91	615.01	708.33	816.72	892.51

Table 5. The comprehensive validity degree of the eleven forecast models based on the real and forecast power consumptions in 1998 and 2001–2005.

Forecast Model	Comprehensive Validity Degree (%)
FM_1	81.09
FM_2	77.90
FM_3	75.56
FM_4	84.32
FM_5	85.80
FM_6	79.03
FM_7	74.87
FM_8	88.91
FM_9	87.23
FM_{10}	78.45
FM_{11}	82.91

Table 6. The results of FMWD-PSO, FMWD-IA-PSO and FMWD-improved-IA-PSO.

Algorithm	Iteration Number	The Forecast Model Weight				
		SFM_1	SFM_2	SFM_3	SFM_4	SFM_5
FMWD-PSO	623	0.0611	0.2120	0.3876	0.0861	0.2532
FMWD-IA-PSO	490	0.2598	0.1662	0.3343	0.0343	0.2054
FMWD-improved-IA-PSO	193	0.4105	0.0401	0.4895	0.0002	0.0597

Using the forecast model weights shown in Table 6, the results of weighted combination forecast (WCF) based on FMWD-PSO, FMWD-IA-PSO and FMWD-improved-IA-PSO, which are abbreviated as WCF-FMWD-PSO, WCF-FMWD-IA-PSO and WCF-FMWD-improved-IA-PSO, are shown in Table 7.

Table 7. The results of weighted combination forecast methods (10^8 kWh).

Weighted Combination Forecast	1998	2001	2002	2003	2004	2005
WCF-FMWD-PSO	434.82	558.22	631.93	720.71	821.93	923.03
WCF-FMWD-IA-PSO	436.28	556.97	630.82	721.19	826.19	936.41
WCF-FMWD-improved-IA-PSO	437.05	556.52	630.98	722.97	831.81	954.96

Four synthesized forecast methods reported by Kang *et al.* [55] are as follows:

(1) Equal weight method: the weights of forecast models are equal, so the combination forecast value of the forecast year c is as follows:

$$y_c = \frac{1}{n_{fm}} \sum_{d=1}^{n_{fm}} y_{c,d}$$
(38)

where $c = n_{his} + 1, n_{his} + 2, ..., n_{his} + n_{fo}, d = 1, 2, ..., n_{fm}$. It is a simple combination forecast method, and both the precision of single forecast model and the relationship between different forecast models are considered.

(2) Variance analysis method: the combination forecast value of the forecast year c is as follows:

$$y_c = \sum_{d=1}^{n_{fm}} \omega_d \cdot y_{c,d}$$
(39)

where $c = n_{his} + 1, n_{his} + 2, ..., n_{his} + n_{fo}, d = 1, 2, ..., n_{fm}$. All forecast models are independent of each other, so the variance of combination forecast can be expressed as follows:

$$Var = \sum_{d=1}^{n_{fm}} \omega_d^2 \cdot \delta_{dd}$$
(40)

where $c = n_{his} + 1, n_{his} + 2, ..., n_{his} + n_{fo}, d = 1, 2, ..., n_{fm}, \delta_{dd}$ is the variance of the forecast model FM_d. To obtain the minimum value of Var on ω_d, the Lagrange multiplier method is used and the weight of FM_d is defined as follows:

$$\omega_d = \cfrac{1}{\delta_{dd}\left(\cfrac{1}{\delta_{11}} + \cfrac{1}{\delta_{22}} + \ldots + \cfrac{1}{\delta_{n_{fm}n_{fm}}}\right)} \tag{41}$$

(3) Optimum fitting method

The forecast model weight determination of the optimum fitting method is based on the deviations of all single forecast models and the complementarily between different forecast models. The deviations of the forecast model FM_d can be expressed as follows:

$$Dev_d = \frac{1}{2}\left(\frac{1}{n_{shis}}\left|\sum_{e=1}^{n_{shis}}(y'-y'_{e,d})\right| + \frac{1}{n_{shis}}\sum_{e=1}^{n_{shis}}\left|y'-y'_{e,d}\right|\right) \tag{42}$$

Therefore, the weight of FM_d is defined as follows:

$$\omega_d = \cfrac{\max\limits_{1\leqslant d'\leqslant n_{fm}} Dev_{d'} + \min\limits_{1\leqslant d'\leqslant n_{fm}} Dev_{d'} - Dev_d}{\sum\limits_{d=1}^{n_{fm}}\left(\max\limits_{1\leqslant d'\leqslant n_{fm}} Dev_{d'} + \min\limits_{1\leqslant d'\leqslant n_{fm}} Dev_{d'} - Dev_d\right)} \tag{43}$$

(4) Optimum forecast method

In this method, n_{fm} forecast models are used to carry out the forecasting respectively, and then these models are compared according to standard deviations, fitting goodness, correlation degree or relative error et al. Lastly the best one is chosen as the final forecast model.

Percentage error (PE) and mean absolute percentage error (MAPE) are used as evaluating indicators to compare the proposed method (WCF-FMWD-improved-IA-PSO) to WCF-FMWD-PSO, WCF-FMWD-IA-PSO, the single forecast models (SFM_1–SFM_5) and the four synthesized forecast methods [55] (equal weight method, variance analysis method, optimum fitting method and optimum forecast method). They are shown in Table 8.

Table 8. The percentage error of the proposed method (WCF-FMWD-improved-IA-PSO) and others.

Forecast Method	Mean Absolute Percentage Error (%)	Percentage Error (%)					
		1998	2001	2002	2003	2004	2005
SFM_1	0.42	−0.22	−0.14	0.43	−0.21	0.03	1.48
SFM_2	1.12	0.02	1.70	2.10	0.25	1.17	−1.51
SFM_3	0.40	−0.18	−0.14	0.44	−0.19	0.02	1.45
SFM_4	6.31	−7.09	4.56	4.77	0.02	−6.91	−14.51
SFM_5	2.41	0	−2.27	−2.20	−2.33	−1.96	−5.69
Equal weight method [55]	0.98	0.04	1.56	1.10	0.80	−1.02	−1.33
Variance analysis method [55]	1.25	0.02	−1.27	−0.20	−2.21	−1.16	−2.63
Optimum fitting method [55]	2.78	−2.09	5.56	0.77	1.02	−2.91	−4.32
Optimum forecast method [55]	1.16	−0.59	2.11	0.97	0.56	1.21	1.50
WCF-FMWD-PSO	0.93	−0.69	0.12	0.49	−0.62	−1.33	−2.36
WCF-FMWD-IA-PSO	0.53	−0.36	−0.11	0.32	−0.55	−0.82	−1.05
WCF-FMWD-improved-IA-PSO	0.35	−0.18	−0.19	0.34	−0.31	−0.14	0.91

From the comparison of PE and MAPE of the proposed method (WCF-FMWD-improved-IA-PSO) and others shown in Figure 4 and Figure 5 and Table 8, we can see the follows:

- The maximum and minimum PE of the single forecast models (SFM_1–SFM_5) are −14.51% and 0 respectively, the maximum and minimum MAPE of the single forecast models (SFM_1–SFM_5) are 6.31% and 0.40% respectively.
- The maximum and minimum PE of the four synthesized forecast methods in Reference [55] are 5.56% and 0.02% respectively, the maximum and minimum MAPE of the four synthesized forecast methods in Reference [55] are 2.78% and 0.98% respectively.
- The maximum and minimum PE of WCF-FMWD-PSO are −2.36% and 0.12% respectively, the MAPE of WCF-FMWD-PSO is 0.93% and the iteration number is 623.
- The maximum and minimum PE of WCF-FMWD-IA-PSO are 1.05% and −0.11% respectively, the MAPE of WCF-FMWD-IA-PSO is 0.53% and the iteration number is 490.
- The maximum and minimum PE of WCF-FMWD-improved-IA-PSO are 0.91% and −0.14% respectively, the MAPE of WCF-FMWD-improved-IA-PSO is 0.35% and the iteration number is 193.

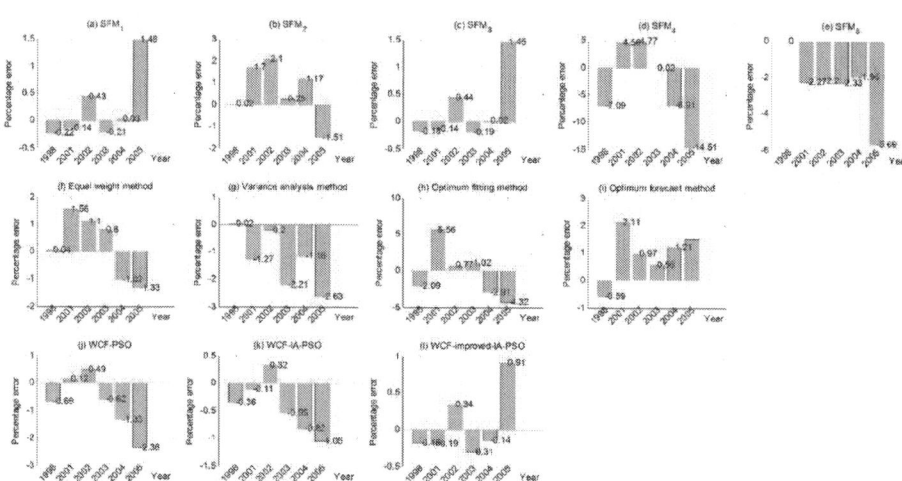

Figure 4. The comparison of PE of the proposed method (WCF-FMWD-improved-IA-PSO) and others.

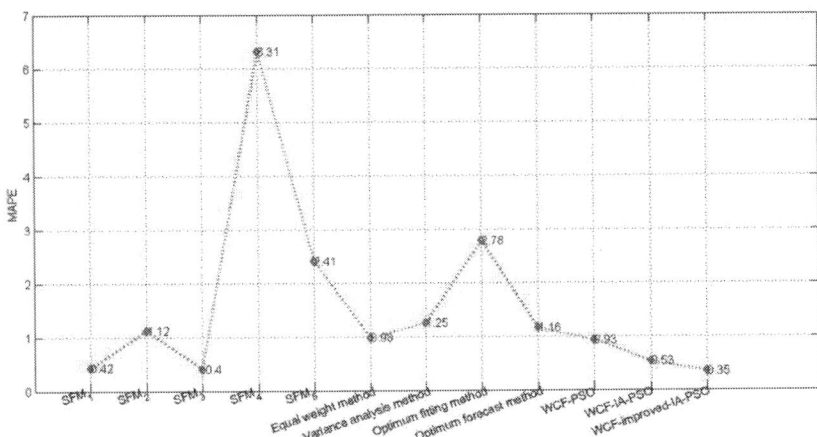

Figure 5. The comparison of MAPE of the proposed method (WCF-FMWD-improved-IA-PSO) and others.

Through the above analysis and the comparison shown in Figure 4 and Figure 5, the proposed method WCF-FMWD-improved-IA-PSO results in an improved accuracy overall (MAPE is 0.35, maximum and minimum PE are 0.91% and −0.14%) when compared against other described methods over the chosen time period and year characteristics. In addition, the proposed method WCF-FMWD-improved-IA-PSO has the fastest convergence rate in forecast model weight determination when compared against WCF-FMWD-PSO and WCF-FMWD-IA-PSO. Therefore, using the proposed method to forecast the medium-and-long-term load is better than using other methods. The correctness and feasibility of the proposed method are proven.

6. CONCLUSIONS

In this paper, we have proposed a robust weighted combination load forecasting method WCF-FMWD-improved-IA-PSO based on forecast model filtering and adaptive variable weight determination to forecast the annual electric load. The contribution and novelty are mainly as follows:

(1) Due to the fact that the forecast year's true load is unknown, the comprehensive validity degree of forecast model is defined by the integration of fitted value relative error and forecast value relative error, and then forecast models are filtered based on their comprehensive validity degrees.

- The definition of validity degree can effectively overcome the inherent shortcomings of error theory. Entirely investigating the fitting level and the validity of forecast model, the comprehensive validity degree definition and the forecast model filtering method can improve the robustness of combination forecasting.

- Revealing the transition pattern between the natural precision and validity degree, the forecast precision estimation method based on Markov chain and cloud model can provide an important basis for the subsequent weighted combination forecasting. In the forecast models' filtering, the better ones will be selected and the worse ones will be eliminated. It can also improve the robustness of combination forecasting.

(2) The improved IA-PSO is used to determine the forecast model weight in combination forecasting. Based on the uniting of immune system's specific information processing mechanism and PSO's global convergence ability, disturbance variable and particle searching speed's adaptive adjustable strategy are introduced to improve the algorithm performance. The particles' diversity is ensured while the convergence speed is increased. It can avoid the local optimal and improve the accuracy.

As can be seen from the case study, the maximum and minimum of percentage error by the proposed method WCF-FMWD-improved-IA-PSO are 0.91%, −0.14% and the mean absolute percentage error is 0.35%, which are smaller than those by the single forecast models (SFM_1–SFM_5), the four synthesized forecast methods [55], WCF-FMWD-PSO and WCF-FMWD-IA-PSO. These indicate that the proposed method has significant superiority over other methods in the terms of annual electric load forecasting accuracy. The iteration number of FMWD-improved-IA-PSO in the proposed method (193) is far less than the iteration numbers of FMWD-PSO and FMWD-IA-PSO (623 and 490), so its advantage that the global optimal solution is reached faster than PSO and IA-PSO is confirmed. In conclusion, the proposed method WCF-FMWD-improved-IA-PSO has a higher robustness and better accuracy, and it can meet the requirements of the annual electric load forecast and can also be applied in the forecast of related fields. In the forecast with analogous features, the proposed method WCF-FMWD-improved-IA-PSO can also be applied.

ACKNOWLEDGMENTS

This work is supported by the National Natural Science Foundation of China (Nos. 61162005 and 61163002), External-Planned Task of State Key Laboratory of Robotics and System, Harbin Institute of Technology (HIT) (No. SKLRS-2013-MS-05), Natural Science Foundation of Ningxia (No. NZ15103), Key Scientific Research Project of Beifang University of Nationalities (No. 2015KJ08), Science & Technology Research Project of Ningxia High School (No. NGY2015149), and Introduction Talent Starting Scientific Research Project of Beifang University of Nationalities for Lianhui Li and Shaohu Ding.

AUTHOR CONTRIBUTIONS

Lianhui Li and Chunyang Mu are the principal investigators of this work. They proposed the robust weighted combination load forecasting method and wrote the manuscript. Shaohu Ding designed the case study solution and checked the whole manuscript. Zheng Wang processed the data in the case study. Runyang Mo and Yongfeng Song did the data analysis work.

NOMENCLATURE

n	number
CH	year characteristic
CHQ	year characteristic quantity
HY	history year
FY	forecast year
CSI	Cosine similarity
FM	forecast model
y'	similar year load
y	forecast year load
y'''	forecast year load by a forecast model
RE'	fitted value relative error
RE	forecast value relative error
P'	fitted precision
P	forecast precision
FIV	fitted validity degree
FOV	forecast validity degree

S	sub-interval
OC	occurrence number
TRN	transition number
TP	transition probability
TM	state transition matrix
IV	initial vector
SM	state matrix
CVE	column vector
Ex	expectation
En	entropy
He	hyper-entropy
En'	normal random number
CV	comprehensive validity degree
SFM	selected forecast model
PO	particle swarm
PA	particle
x	position
v	speed
LF	learning factor
rand	random number

w	inertia coefficient
PRO	probability
AF	affinity
CON	concentration
iter	iteration
F	fitness function
PF	particle fitness
r	vaccination times
Var	variance
Dev	deviation
Greek letters	
σ	standard deviation
α	empirical coefficient
χ	weight
β	disturbance variable
Δ	swarm convergence degree
η	coefficient used to control the upper limit
ω	forecast model's weight

Superscripts

q	the gth sub interval
h	the hth sub-interval
(h, g)	the transition from the hth to the gth
(k)	the kth iteration
(q)	step
q	the qth power

Subscripts

his	history year
fo	forecast year
$shis$	selected history year
fm	forecast model
sfm	selected forecast model
si	sub-interval
pa	particle
apa	added particle
avg	average
AVG	the average of the numbers which are bigger than the global average
a	the ath year characteristic

b	the bth year characteristic quantity
c	the cth forecast year
d	the dth forecast model
e	the eth similar year
i	the ith column vector
j	the jth selected forecast model
l	the lth particle
m	the dimensional number of solution space
t	the tth element

REFERENCES

1. Kouhi, S.; Keynia, F. A new cascade NN based method to short-term load forecast in deregulated electricity market.*Energy Convers. Manag.* **2013**, *71*, 76–83.
2. Chaturvedi, D.K.; Sinha, A.P.; Malik, O.P. Short term load forecast using fuzzy logic and wavelet transform integrated generalized neural network. *Int. J. Electr. Power Energy Syst.* **2015**, *67*, 230–237.
3. Yang, Z.C. Electric load evaluation and forecast based on the elliptic orbit algorithmic model. *Int. J. Electr. Power Energy Syst.* **2012**, *42*, 560–567.
4. Bennett, C.; Stewart, R.A.; Lu, J. Autoregressive with exogenous variables and neural network short-term load forecast models for residential low voltage distribution networks. *Energies* **2014**, *7*, 2938–2960.
5. Li, F.; Buxiang, Z.; Shi, C. The Medium and Long Term Load Forecasting Combined Model Considering Weight Scale Method. *TELKOMNIKA Indones. J. Electr. Eng.* **2013**, *11*, 2181–2186.
6. Li, R.; Su, H.; Wang, Z. Medium-and long-term load forecasting based on heuristic least square support vector machine. *Power Syst. Technol.* **2011**, *35*, 195–199. (In Chinese).

7. Mello, P.E.; Lu, N.; Makarov, Y. An optimized autoregressive forecast error generator for wind and load uncertainty study. *Wind Energy* **2011**, *14*, 967–976.

8. Li, H.; Guo, S.; Zhao, H. Annual electric load forecasting by a least squares support vector machine with a fruit fly optimization algorithm. *Energies* **2012**, *5*, 4430–4445.

9. .; Gerek, Ö.N.; Kurban, M. A novel modeling approach for hourly forecasting of long-term electric energy demand. *Energy Convers. Manag.* **2011**, *52*, 199–211.

10. Suganthi, L.; Samuel, A.A. Energy models for demand forecasting—A review. *Renew. Sustain. Energy Rev.* **2012**, *16*, 1223–1240.

11. Sheikh, S.K.; Unde, M.G. Short Term Load Forecasting using ANN Technique. *Int. J. Eng. Sci. Emerg. Technol.* **2012**, *1*, 97–107.

12. Wu, Y. The medium and long-term load forecasting based on improved D-S evidential theory. *Trans. China Electrotech. Soc.* **2012**, *27*, 157–162. (In Chinese).

13. Long, R.; Mao, Y.; Mao, L. A combination model for medium-and long-term load forecasting based on induced ordered weighted averaging operator and Markov chain. *Power Syst. Technol.* **2010**, *3*, 150–156. (In Chinese).

14. Chen, H.Y. *The Validity Theory of Combination Forecasting Method and Its Application*; Science Press: Beijing, China, 2008; pp. 76–109. (In Chinese)

15. Chen, H.Y. Research on combination forecasting model based on effective measure of forecasting methods. *Forecasting* **2001**, *20*, 72–73.

16. Sun, G.Q.; Yao, J.G.; Xie, Y.X. Combination forecast of medium-and-long-term load using fuzzy adaptive variable weight based on fresh degree function and forecasting availability. *Power Syst. Technol.* **2009**, *33*, 103–107. (In Chinese).

17. Chen, C.; Guo, W.; Fan, J.Z. Combined method of mid-long term load forecast based on improved forecasting effectiveness. *Relay* **2007**, *35*, 70–74. (In Chinese).

18. Jin, X.; Luo, D.S.; Sun, G.Q. Sifting and combination method of medium-and-long-term load forecasting model. *Proc. Chin. Soc. Univ. Electr. Power Syst. Autom.* **2012**, *24*, 150–156. (In Chinese).

19. Hernandez, L.; Baladrón, C.; Aguiar, J.M. Short-term load forecasting for microgrids based on artificial neural networks. *Energies* **2013**, *6*, 1385–1408.

20. Gofman, A.V.; Vedernikov, A.S.; Vedernikova, E.S. Increasing the accuracy of the short-term and operational prediction of the load of a power system using an artificial neural network. *Power Technol. Eng.* **2013**, *46*, 410–415.

21. Li, H.; Guo, S.; Li, C. A hybrid annual power load forecasting model based on generalized regression neural network with fruit fly optimization algorithm. *Knowledge-Based Syst.* **2013**, *37*, 378–387.

22. Paparoditis, E.; Sapatinas, T. Short-term load forecasting: The similar shape functional time-series predictor. *IEEE Trans. Power Syst.* **2013**, *28*, 3818–3825.

23. Weron, R.; Taylor, J. Discussion on "Electrical load forecasting by exponential smoothing with covariates". *Appl. Stoch. Models Bus. Ind.* **2013**, *29*, 648–651.

24. Li, D.C.; Chang, C.J.; Chen, C.C. Forecasting short-term electricity consumption using the adaptive grey-based approach—An Asian case. *Omega* **2012**, *40*, 767–773.

25. Ismail, N.A.; King, M. Factors influencing the alignment of accounting information systems in small and medium sized Malaysian manufacturing firms. *J. Inf. Syst. Small Bus.* **2014**, *1*, 1–20.

26. Kodogiannis, V.S.; Amina, M.; Petrounias, I. A clustering-based fuzzy wavelet neural network model for short-term load forecasting. *Int. J. Neural Syst.* **2013**, *23*, 1557–1565.

27. Hong, T.; Wang, P. Fuzzy interaction regression for short term load forecasting. *Fuzzy Optim. Decis. Mak.* **2014**, *13*, 91–103.

28. Borges, C.E.; Penya, Y.K.; Fernandez, I. Evaluating combined load forecasting in large power systems and smart grids.*IEEE Trans. Ind. Inf.* **2013**, *9*, 1570–1577.

29. Che, J.; Wang, J.; Wang, G. An adaptive fuzzy combination model based on self-organizing map and support vector regression for electric load forecasting. *Energy* **2012**, *37*, 657–664.

30. Liu, Z.; Li, W.; Sun, W. A novel method of short-term load forecasting based on multiwavelet transform and multiple neural networks. *Neural Comput. Appl.* **2013**, *22*, 271–277.

31. Wang, J.; Wang, J.; Li, Y. Techniques of applying wavelet de-noising into a combined model for short-term load forecasting. *Int. J. Electr. Power Energy Syst.* **2014**, *62*, 816–824.

32. Enayatifar, R.; Sadaei, H.J.; Abdullah, A.H. Imperialist competitive algorithm combined with refined high-order weighted fuzzy time series

(RHWFTS-ICA) for short term load forecasting. *Energy Convers. Manag.* **2013**, *76*, 1104–1116.

33. Ko, C.N.; Lee, C.M. Short-term load forecasting using SVR (support vector regression)-based radial basis function neural network with dual extended Kalman filter. *Energy* **2013**, *49*, 413–422.

34. Ma, S.; Chen, X.; Liao, Y.; Wang, G.; Ding, X.; Chen, K. The variable weight combination load forecasting based on grey model and semiparametric regression model. In Proceedings of the TENCON 2013— 2013 IEEE Region 10 Conference (31194), Xi'an, China, 22–25 October 2013; pp. 1–4.

35. Niu, D.; Shi, H.; Wu, D.D. Short-term load forecasting using bayesian neural networks learned by Hybrid Monte Carlo algorithm. *Appl. Soft Comput.* **2012**, *12*, 1822–1827.

36. P.; Dai, J. Optimization selection of correlative factors for long-term load prediction of electric power. *Comput. Simul.* **2008**, *5*, 226–229.

37. Wang, Q.; Wang, Y.L.; Zhang, L.Z. An approach to allocate impersonal weights of factors influencing electric power demand forecasting. *Power Syst. Technol.* **2008**, *32*, 82–86. (In Chinese).

38. Lei, S.L.; Gu, L.; Yang, J.; Liu, X.Y. Analysis of electric power load characteristics and its influencing factors in chongqing region. *Electr. Power* **2014**, *12*, 61–71. (In Chinese).

39. Liao, F.; Congying, X.U.; Yao, J.; Cai, J.; Chen, S. Load characteristics of change region and analysis on its influencing factors. *Power Syst. Technol.* **2012**, *7*, 117–125.

40. Karakoc, E.; Cherkasov, A.; Sahinalp, S.C. Distance based algorithms for small biomolecule classification and structural similarity search. *Bioinformatics* **2006**, *14*, 243–251.

41. Ye, J. Cosine similarity measures for intuitionistic fuzzy sets and their applications. *Math. Comput. Model.* **2011**, *1*, 91–97.

42. Zhu, S.; Wu, J.; Xiong, H.; Xia, G. Scaling up top-K cosine similarity search. *Data Knowl. Eng.* **2011**, *1*, 60–83.

43. Keilson, J. *Markov Chain Models—Rarity and Exponentiality*; Springer Science and Business Media: Berlin, Germany, 2012.

44. Yoder, M.; Hering, A.S.; Navidi, W.C.; Larson, K. Short-term forecasting of categorical changes in wind power with Markov chain models. *Wind Energy* **2014**, *17*, 1425–1439.

45. Li, D.Y.; Liu, C.Y. Study on the universality of the normal cloud model. *Eng. Sci.* **2004**, *6*, 28–34.

46. Wang, G.; Xu, C.; Li, D. Generic normal cloud model. *Inf. Sci.* **2014**, *280*, 1–15.
47. Kennedy, J. Particle swarm optimization. In *Encyclopedia of Machine Learning*; Springer Publishing Company: New York, NY, USA, 2010; pp. 760–766.
48. Zhao, F.; Li, G.; Yang, C.; Abraham, A.; Liu, H. A human-computer cooperative particle swarm optimization based immune algorithm for layout design. *Neurocomputing* **2014**, *132*, 68–78.
49. Fu, X.; Li, A.; Wang, L.; Ji, C. Short-term scheduling of cascade reservoirs using an immune algorithm-based particle swarm optimization. *Comput. Math. Appl.* **2011**, *6*, 2463–2471.
50. Afshinmanesh, F.; Marandi, A.; Rahimi-Kian, A. A novel binary particle swarm optimization method using artificial immune system. In Proceedings of the International Conference on Computer as a Tool, EUROCON 2005, Belgrade, Serbia, 21–24 November 2005; Volume 1, pp. 217–220.
51. Wu, J.M.; Zuo, H.F.; Chen, Y. A combined forecasting method based on particle swarm optimization with immunity algorithms. *Syst. Eng. Theory Method Appl.* **2006**, *15*, 229–233. (In Chinese).
52. China National Bureau of Statistics. *China Energy Statistical Yearbook*; China Statistics Press: Beijing, China, 2011. (In Chinese).
53. Deng, J.L. Introduction to grey system theory. *J. Grey Syst.* **1989**, *1*, 1–24. (In Chinese).
54. Gardner, E.S. Exponential smoothing: The state of the art. *J. Forecast.* **1985**, *1*, 1–28.
55. Kang, C.Q.; Xia, Q.; Liu, M. *Power System Load Forecasting*; China Electric Power Press: Beijing, China, 2007; pp. 73–75. (In Chinese)

Chapter 3

Gas Turbine Transient Performance Tracking Using Data Fusion Based on an Adaptive Particle Filter

Feng Lu [1,2,*], Yafan Wang [1], Jinquan Huang [1,2,*] and Yihuan Huang [1]

[1] *Jiangsu Province Key Laboratory of Aerospace Power Systems, Nanjing University of Aeronautics and Astronautics, Nanjing 210016, China*
[2] *Collaborative Innovation Center of Advanced Aero-Engine, Beijing 100191, China*

ABSTRACT

This paper considers the problem of gas turbine transient performance tracking in a cluttered environment. To increase the accuracy and robustness of state estimation, a data-fusion nonlinear estimation method based on an adaptive particle filter (PF) is proposed. This method needs local estimates transmitted to a central filtering unit for data fusion, and then global data feedback to the local PF for consensus propagation. The computational burden is shared by the local PF and central filtering unit in the data-fusion architecture. Furthermore, the PF algorithm used for the data fusion is embedded with the prior knowledge of engine health condition and adaptive to the measurement noise, and hence is called the adaptive PF. The heuristic information of state variables represented by inequality constraints tunes the local estimates by a probability density truncation method. The covariance of measurement noise is calculated by wavelet transform and utilized to update the particle importance function of the real time PF. The performance improvements of the proposed method are indicated through extensive experiments for gradual and abrupt shift performance tracking under conditions of gas turbine transient operation.

Keywords: gas turbine; performance tracking; data fusion; particle filter (PF); probability density truncation; wavelet transform

1 INTRODUCTION

Gas turbine engines provide the power for airplanes, and their reliability is vital to flight safety and performance. Nevertheless, the engine working conditions are terrible and they usually must endure high speeds, extreme temperatures and strong vibrations. Erosion and fouling of major components are unavoidable and result in a gradual deterioration of engine performance during its lifetime. Besides, foreign/domestic object damage will cause engine performance to sharply shift, and it is called an abrupt fault [1]. Engine data are periodically collected by airlines for engine health evaluation. Maintenance schedules adapted to reliable health tracking results leads to safe operation and reduced costs [2]. Hence, how to get reliable information about engine health conditions in time has drawn a lot of attention.

Engine health parameters, which are correction factors for the efficiency of the major components, are introduced to quantify the engine performance changes [3]. The engine performance tracking problem can be regarded as calculating the health parameters. Various methods such as Kalman filters (KFs), neural networks, fuzzy logic, genetic algorithms and expert systems have been proposed to obtain health parameters for engine health monitoring [3,4,5,6,7]. KF-based methods seem to be the most common ones for gas turbine health estimation, but these techniques are mainly focused on engine steady state, e.g., under cruise or average conditions [8].

The engine transient condition represents how a gas turbine operates from one steady state to another steady state when the input changes. The rotating components of the engine (fan, compressor, high-pressure turbine (HPT) and low-pressure turbines (LPT)) run close to their surge boundaries during transient behavior, and some performance parameters exceed their extreme value in a short time. Generally speaking, a low-bypass engine performance anomaly more easily occurs during a transient process, e.g., acceleration and deceleration. That is to say, gas turbine transient performance tracking is more urgent compared to steady state monitoring, but up to this point in time no studies of dynamic behavior monitoring for gas turbines have been presented. This paper proposes an adaptive particle filter (PF) based data fusion approach to engine transient health monitoring.

In this paper we emphasize the problem of monitoring the health performance of key gas turbine rotating components, including fans, compressors (high-pressure compressors (HPCs)), high-pressure turbines (HPTs) and low-pressure turbines (LPTs). Failures of engine actuators, sensors, and other components are not considered. The engine performance tracking of gradual drifts and abrupt shifts during transient operation is mainly addressed, but the technique can also be applied to engine steady behavior. A data fusion architecture to monitor engine health conditions is designed, and prior knowledge of engine health state and measurement noise is utilized for a Monte Carlo simulation to tune the state estimates. The present contribution is a derivation of a data fusion method based on adaptive PF for gas turbine transient performance tracking at the theoretical, implementation and performance level.

This paper is organized as follows: Section 2 presents a review of the basic PF and the problem formulation; the adaptive PF algorithm and data fusion architecture implementation for gas turbine health monitoring is given in Section 3; the performance comparisons of the proposed method for gradual and sharp shifts in different operation conditions are discussed in Section 4; and Section 5 concludes this paper.

2. PROBLEM FORMULATION AND PARTICLE FILTER

2.1. Problem Formulation

A gas turbine engine is a low-bypass turbofan engine, see Figure 1. A single inlet supplies airflow to the fan. Air leaving the fan is separated into two streams: one stream passes through the engine core, and the other stream passes through the annular bypass duct and then leaves. The fan is driven by the low pressure turbine. The air passing through the engine core moves through the HPC, which is driven by the HPT. Fuel is injected in the combustor and burned to produce hot gas for driving the turbines. The gas leaves the LPT and is mixed with the air from the bypass duct through the convergent nozzle, which has a variable cross section area.

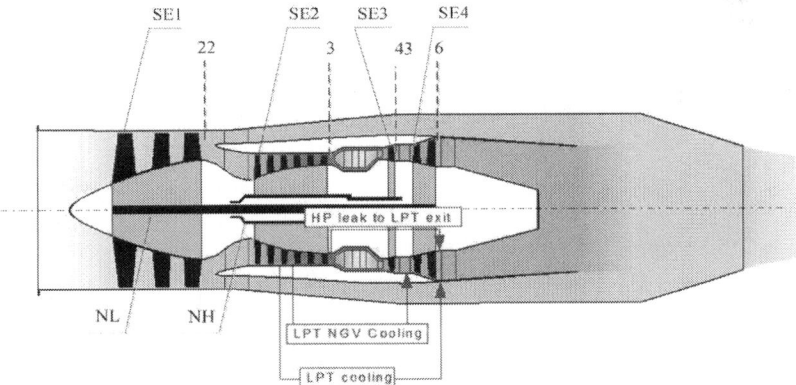

Figure 1. Schematic representation of a gas turbine engine.

Considering air flow mass, power and momentum conservation laws [9,10], a nonlinear aero-thermodynamic model of a gas turbine engine is given by:

$$x_{k+1} = f(x_k, u_k) + w_k$$

$$y_k = h(x_k, u_k) + v_k \tag{1}$$

where k is the time index, y is the 8-element measured output, x is the 6-element augmented state, and u is the 2-element control input. The noise terms wk and vk represent the process inaccuracies and measurement inaccuracies in the model. The sensor measurements are low-pressure spool speed N_L, high-pressure spool speed N_H, fan outlet pressure P_{22}, HPC outlet pressure P_3, fan outlet temperature T_{22}, HPC outlet temperature T_3, HPT outlet temperature T_{43} and LPT outlet temperature T_6. The augmented state x includes the 2-element original state x_o (N_L and N_H) and health parameter vector p(fan efficiency $SE1$, HPC efficiency $SE2$, HPT efficiency $SE3$ and LPT efficiency $SE4$, where section efficiency is defined SE). The elements of the control vector in the model are fuel flow W_f and nozzle area A_8, which defines the engine operating point. There are two information entropy definitions used to select the system parameters. Auto Information Entropy is utilized to select the measured parameters, while the Cross Information Entropy to analyze the correlations between control, measured and health parameters.

In the framework of gas turbine performance monitoring, the quantities of interest are the differences between the estimated engine health parameters status and their reference ones. The actual engine performance is represented by the estimated values of health parameter, the prior value of which is adapted to the current measurements in a recursive approach. The block diagram of the model-based approach to monitoring the engine health condition is shown in Figure 2, and it is a closed-loop state estimator correcting structure. The KF, especially the linear KF (LKF), is an optimal state estimator for linear systems with noisy and inaccurate measurements, and is widely used for gas turbine engine health monitoring [11].

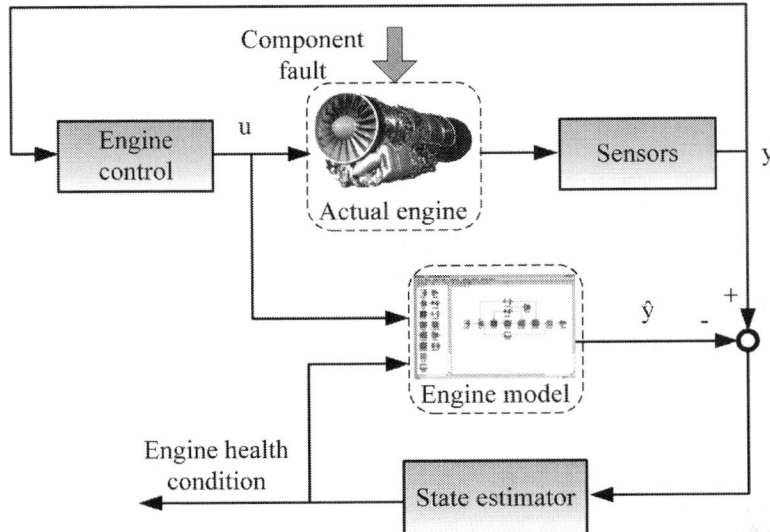

Figure 2. Gas turbine health monitoring based on a state estimator.

The state variable model (SVM) is a piecewise linear representation of the engine and needs to be established before the LKF implementation. It should be pointed out that the SVM obtained at a steady operating point is only a representation of the engine near this operating point. The state estimation accuracy by the LKF varies with the distance from the actual operating point to the design point, and also with the magnitude of the state deviation [12]. That means that the strong nonlinearity of gas turbine makes the LKF effective only in a small working range around the nominal state, and it can't do well in cases of transient behavior, e.g., during acceleration and deceleration.

Note that state variables change together in a situation of normal gradual engine deterioration, and the SVM is derived by the partial derivative calculation in the linearization process. Thus it is impossible for the SVM to correctly describe simultaneous shifts of all states, thus the tracking accuracy of gradual shifts by the LKF is not satisfactory. In addition, the assumption that the derivative of health parameter h is approximated by zero is utilized in the LKF equations, which is because engine deterioration generally occurs very slowly relative to the dynamics of the original state variables x_0. Therefore any sharp shift of health parameters is not well estimated by the LKF due to the contradiction with the hypothesis $h'=0$. Hence, the generic LKF-based engine monitoring approach to the abrupt and multi-state gradual shift has a theoretical drawback.

To aim at the above issues, various nonlinear filtering methods are proposed and applied to state estimation, especially the extended Kalman filter (EKF), unscented Kalman filter (UKF) and PF [13,14]. Nevertheless, the EKF method is often used for weak nonlinear Gaussian systems due to a Taylor Series expansion truncated to the first order [8], and the performance closely depends on how often Jacobians are updated. Compared to the EKF and UKF, the PF is based on sequential Monte Carlo sampling theory, and it does not necessitate simplification of nonlinearity or any hypothesis of specific distributions [15,16]. Therefore, the PF-based approach is the one of the best ways that we address strongly nonlinear engine health monitoring issues in the following section.

2.2. The Particle Filter

The nonlinear model of a gas turbine is given by Equtaion (1), and let $x_{0:k} \triangleq \{x_0, \cdots, x_k\}$ and $y_{1:k} \triangleq \{y_1, \cdots, y_k\}$ denote the series of the state and measurement. Assume the probability density function (PDF) of the prior condition is p(x_0), and the posterior $\text{PDF } p(x_{0:k}|y_{1:k})$ is characterized by a set of weighted random samples $\{x_{0:k}^i, w_k^i\}_{i=1}^N$, wherein N_s is the particle number. The particle set $\{x_{0:k}^i; i = 0, \cdots, N\}$ is associated to the weights $\{w_k^i; i = 0, \cdots, N\}$, and the PDF at time k can be approximated by:

$$p(x_{0:k}|y_{1:k}) \approx \sum_{i=1}^N w_k^i \delta(x_{0:k} - x_{0:k}^i)$$

$$\sum_{i=1}^N w_k^i = 1$$

(2)

The case that the particles for Monte Carlo sampling are directly generated from the posterior PDF $p(x_{0:k}|y_{1:k})$ is expected, but it is generally unavailable. Thus the importance sampling distribution function $q(x_{0:k}|y_{1:k})$ is defined before sampling:

$$q(x_k|x_{0:k-1}, z_{1:k}) = q(x_k|x_{k-1}, z_k)$$

$$q(x_k^i|x_{k-1}^i, z_k) = p(x_k^i|x_{k-1}^i)$$

(3)

Then the ith particle weight wik can be approximated by:

$$w_k^i \propto \frac{p(x_{0:k}|y_{1:k})}{q(x_{0:k}|y_{1:k})} \propto \frac{p(y_k|x_k^i)p(x_k^i|x_{k-1}^i)p(x_{0:k-1}^i|y_{1:k-1})}{q(x_k^i|x_{0:k-1}^i,y_{1:k})q(x_{0:k-1}^i|y_{1:k-1})} = w_{k-1}^i \frac{p(y_k|x_k^i)p(x_k^i|x_{k-1}^i)}{q(x_k^i|x_{0:k-1}^i,y_{1:k})}$$

(4)

With this choice and normalization, the importance weights can be computed:

$$w_k^i \propto w_{k-1}^i p(y_k|x_k^i)$$

$$w_k^i = w_k^i / \sum_{i=1}^{N} w_k^i$$

(5)

One problem of the basic PF algorithm is that more particles have negligible weights after a few recursive steps. This implies that particle degeneracy occurs and a large computational effort for updating particle is meaningless. A typical method for solving this issue is importance re-sampling, and each particle is assigned by equal weight $w_k^i = 1/N$ whenever the effective number N_{eff} of particles becomes less than a threshold value Nth.

$$N_{eff} = \frac{1}{\sum_{i=1}^{N}(w_k^i)^2} < N_{th}$$

(6)

Once Nth is close to value $N_{eff,k}$, all particles have almost the same significance.

The architecture of a conventional PF mentioned above is that the measured data from different sensors are sent to one central PF to process together, and this is so-called the central architecture [15,17]. The advantage of this architecture is minimal information loss, but it also raises the problems that all measurements are identically treated at a time and the central filter bears a heavy computational burden, especially in the framework of the PF [18,19,20]. With the development of information fusion, the fusion PF structure employing a bank of local PFs and one master filter is presented and the computational effort is then shared by several filters [21,22].

In the case of the model-based approach to gas turbine health monitoring, the heuristic health knowledge is usually ignored and not considered in estimate algorithms. The typical anomaly modes and gradual deterioration rules of the engine have been summarized, and the magnitude ranges of health parameters are then determined. This prior knowledge about health parameters is represented by the constraints, and it can be used for state estimation in the PF algorithm. Furthermore, measurement noise levels vary with the order of the sensed value during the engine's dynamic operation. Generally speaking, the magnitude of sensed noise increases as the engine works at a larger operating point during the dynamic operation. Hence, it is very important for the PF to tune the estimate with the noise level in the situation of the transient behavior of the engine.

3. ADAPTIVE FUSION PARTICLE FILTER

3.1. The Adaptive Particle Filter

3.1.1. Particle Filter with Inequality Constraints

The prior state information can be described by equality or inequality constraints in the state estimation algorithm. These constraint approaches are mainly concentrated and applied to linear systems, especially combined with a KF, which has been proved to increase performance estimation accuracy. KFs with equality constraint approaches include the moving horizon estimation and smoothly constrained KF [23]. The estimate projection and truncation methods are incorporated to inequality constraints on state estimates, and the latter one has better performance for deterioration tracking due to its handling of two-sided constraints. The nonlinear state estimation with inequality constraints is more useful to gas turbine engine health monitoring, but there are not enough theoretical studies. For a nonlinear dynamic system, the heuristic knowledge represented by inequality constraints is introduced to the PF algorithm. This idea based on a previously published method [24] has been extended to the nonlinear PF approach for abrupt shift tracking.

The prior information we used in this paper is derived from the gradual deterioration rule and abrupt fault feature. Engine performance degrades with use, and the health parameters never improve and change in one way. For example, we know that the health parameter representing engine efficiency declines over time, and the parameter usually varies within the -10% magnitude due to abrupt faults. With this information, we can determine the bound of the state in advance and it is represented by the inequality constraint of the state estimator. The probability density function (PDF) of the PF estimate at the prior inequality constraint is truncated and the constrained filter estimate as the mean of the truncated PDF is calculated. Now consider the nonlinear system of Equation (1) where the s scalar constraints are added:

$$a_{k,m} \leqslant \phi_{k,m}^{\mathrm{T}} x_k \leqslant b_{k,m}, \qquad m = 1, \cdots, s \qquad (7)$$

where $a_{k,m} \leqslant b_{k,m}, a_{k,m}$ and $b_{k,m}$ are the two sided constraint of the mth health parameter at time k. The health parameter estimate $\hat{x}_k$ and covariance P_k is derived by the unconstraint PF algorithm:

$$\hat{x}_k = \sum_{i=1}^{N} w_k^i x_k^i$$

(8)

$$P_k = \sum_{i=1}^{N} w_k^i (x_k^i - \hat{x}_k)(x_k^i - \hat{x}_k)^{\mathrm{T}}$$

The PF algorithm with inequality constraints is to truncate the unconstrained PDF, $N(x_k, P_k)$. Once the mean $\tilde{x}_k$ and covariance $\tilde{P}_k$ of the truncated PDF are computed, we can obtain the constrained health parameter estimate.

The state estimate $\tilde{x}_{k,m}$ and covariance $\tilde{P}_{k,m}$ are defined after the first m scalar constraints enforced, and then the new transformation is performed:

$$z_{k,m} = S_m W_m^{-\frac{1}{2}} T_m^{\mathrm{T}} (x_k - \tilde{x}_{k,m})$$

$$T_m W_m T_m^{\mathrm{T}} = \tilde{P}_{k,m}$$

(9)

$$S_m W_m^{\frac{1}{2}} T_m^{\mathrm{T}} f_{k,m} = [\; (f_{k,m}^T \tilde{P}_{k,m} f_{k,m}^{\frac{1}{2}}) \quad 0 \quad L \quad 0 \;]^{\mathrm{T}}$$

where T_m is orthogonal, W_m is diagonal (these two quantities, T_m and W_m, can be derived from the Jordan canonical decomposition of $\tilde{P}_{k,m}$) and S_m is obtained using Gram–Schmidt orthogonalisation. The first m inequality constraints are normalized:

$$c_{k,m} \leqslant [1 \quad 0 \quad 0 \cdots 0] z_{k,m} \leqslant d_{k,m}$$

$$c_{k,m} = \frac{a_{k,m} - \phi_{k,m}^{\mathrm{T}} \tilde{x}_{k,m}}{(\phi_{k,m}^{\mathrm{T}} \tilde{P}_{k,m} \phi_{k,m})^{1/2}}$$

(10)

$$d_{k,m} = \frac{b_{k,m} - \phi_{k,m}^{\mathrm{T}} \tilde{x}_{k,m}}{(\phi_{k,m}^{\mathrm{T}} \tilde{P}_{k,m} \phi_{k,m})^{1/2}}$$

Since $z_{k,m}$ is an identity covariance with statistically independent element, only the first element is constrained in Equation (10), and the PDF truncation reduces to a one-dimensional PDF. The part outside of the constraints is removed due to the fact $z_{k,m}$ lays between $c_{k,m}$ and $d_{k,m}$. The truncated PDF is normalized after computing the area of the inside portion of the PDF. The mean μm and variance $\sigma^2 m$ of the first element of $\tilde{z}_{k,m}$ with the constraint enforcement are expressed by:

$$\alpha = \frac{\sqrt{2}}{\sqrt{\pi}(\mathrm{erf}(d_{k,m}/\sqrt{2}) - \mathrm{erf}(c_{k,m}/\sqrt{2}))}$$

$$\mu_m = \alpha \int_{c_{k,m}}^{d_{k,m}} \xi \exp(-\xi^2/2)d\xi = \alpha[\exp(-c_{k,m}^2/2) - \exp(-d_{k,m}^2(k)/2)]$$

$$\sigma^2_m = \alpha \int_{c_{k,m}}^{d_{k,m}} (\xi - \mu_m)^2 \exp(-\xi^2/2)d\xi$$

$$= \alpha[\exp(-c_{k,m}^2/2)(c_{k,m} - 2\mu_m) - \exp(-d_{k,m}^2/2)(d_{k,m} - 2\mu_m)] + \mu_m^2 + 1$$

$$(11)$$

where α is a magnification factor. The inverse transformation of the Equation (9) is taken, and the mean and variance of the constrained state estimate are therefore given as:

$$\tilde{x}_{k,m+1} = T_m W_m^{1/2} S_m^{\mathsf{T}} z_{k,m+1} + \tilde{x}_{k,m}$$

$$\tilde{P}_{k,m+1} = T_m W_m^{1/2} S_m^{\mathsf{T}} \tilde{C}_{k,m+1} S_m W_m^{1/2} T_m^{\mathsf{T}}$$

$$(12)$$

We repeat the process of Equations (9)–(12) to enforce the next constraint to the state estimate and jump out of the iteration until $m = s$. Hence, we obtain the state estimate and covariance at time k as $x_k = \tilde{x}_{k,s}, P_k = \tilde{P}_{k,s}$ by the constrained PF algorithm.

3.1.2. Measure Noise Tuned Particle Filter

As we know, particle importance weight w_i is closely dependent on the distance between the real sensed value and its estimate, and the smaller the distance, the larger the weight. In the basic PF algorithm, the particle importance weight w_i is defined as follows:

$$w_i = \frac{1}{\sqrt{2\pi}\sigma} \exp(-\frac{(y - \hat{y}_i)^2}{2\sigma^2})$$

$$(13)$$

where the quantity σ is consistent with the covariance of measurement noise due to the fact that the sensed noise usually is generally recognized as Gaussian white noise. We set the quantity σ by the heuristic experience of the noise and it remains constant in the conventional PF algorithm. However, the probability distribution of the sensed noise changes in engine transient behavior, and it raises the problem of measurement noise uncertainty with regards to state estimation. The deviation from the actual measurement noise will result in a decline in estimation accuracy. In the paper the on-line tuning quantity σ of the PF with the measurement noise variance with the help of the wavelet transform is designed.

The measured stream yω(k) could be approximated by a low-order polynomial or a piecewise low-order polynomial in an observation interval [25]. This interval size is corresponding to sampling number. Suppose the polynomial of the sensed stream yω(k) is described as follows:

$$y_\omega(k) = a_0 + a_1 k + \cdots + a_M k^M + \omega \tag{14}$$

Then the wavelet transform is implemented to the sensed stream mw(k):

$$W_{y_\omega}(s,\tau) = y_\omega(k) * \varphi_{s,\tau}(k) = W_y(s,\tau) + W_\omega(s,\tau)$$
$$\varphi_{s,\tau}(k) = \frac{1}{\sqrt{s}} \varphi(\frac{k-\tau}{s}) \tag{15}$$

where φ(k) is a wavelet function, "*" is the convolution operation, Wy(s,τ) is an approximate part of the wavelet coefficients and Wω(s,τ) represents a detailed part of the wavelet coefficients. We define a vanishing moment α of the function φ(k), and the measurement noise part is extracted from the sensed stream part once the quantity α is greater than the highest-order of the polynomial, namely α > M. Then the standard deviation of the noisy term in an interval can be calculated by the following expression:

$$\sigma = \frac{1}{M} \text{Med}(|W_{y_\omega}(s, t_h)|) \tag{16}$$

where the scale s equals to 0.5, and the series Wyω(s,th) is the $K/2$ wavelet coefficients of $\{y_\omega(k) | k = 0, 1, \cdots K\}$, wherein $0 \leqslant h \leqslant K/2$. Normalizing parameter M is usually set at 0.6745, and the function Med() represents the middle value calculation of a series.

In order to tune the quantity σ of the PF algorithm in real-time during the engine dynamic behavior, the measurement noise of a series in an interval is computed by a wavelet transformation. The sensed stream yω is segmented by a window with fixed width L, and it slides forward along the sampling time. The variance of the sensed noise part in the sliding window is estimated, therefore, the framework of the adaptive PF algorithm is established as shown in Figure 3.

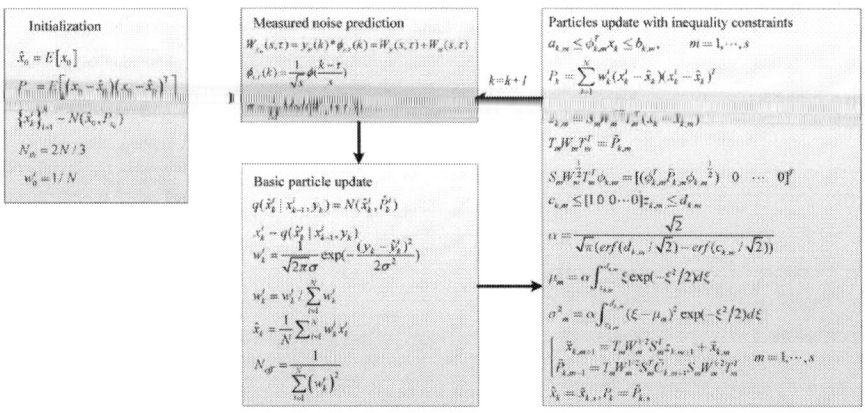

Figure 3. The framework of the adaptive particle filter (PF) algorithm.

3.2. Data Fusion Based on Adaptive Particle Filter

The fusing PF based on the integration of adaptive PF algorithm and information fusion theory is proposed for gas turbine transient performance monitoring. The sensor used to detect engine health condition is divided into several teams, and each team is applied to estimate the local estimates by the local PF. The engine component layout and thermodynamics of operation are taken into account for the partitions of local filters in the fusion architecture. The fan of the engine is a cold and low pressure (LP,) component, the compressor is a cold and high pressure (HP) one, HPT is a hot and HP one, and LPT is a hot and LP one [26,27]. Two different ways to partitioning are performed, namely, the cold-hot component partition and the LP-HP component partition. Therefore, there are four combinations of the four rotating components above, which are the cold group, the hot group, the HP group and the LP group. For example, the cold partition includes two components (fan and compressor). Likewise, the hot partition consists of HPT and LPT, the HP partition of compressor and HPT, and the LP partition of fan and LPT.

For the cold group, the measured parameters are T_{22}, P_{22}, T_3 and P_3. In a similar manner the measurements T_{43} and T_6 are for the hot group, the sensors T_{22}, P_{22} and T_6 for the LP group, and the sensors T_3, P_3, and T_{43} for the HP group. In addition, two spool speeds (N_L and N_H) are important quantities representing engine operation status, which are communal measurements and utilized in each part. Hence, the measurements of four local filter can be denoted by $y_1 = [N_L, N_H, T_{22}, P_{22}, T_3, P_3]$, $y_2 = [N_L, N_H, T_{43}, T_6]$, $y_3 = [N_L, N_H, T_{22}, P_{22}, T_6]$, $y_4 = [N_L, N_H, T_3, P_3, T_{43}]$. Although the measurements of each engine component partition are different, the health parameters to be estimated are the same in every local filter, namely, *SE1*, *SE2*, *SE3* and *SE4*.

The data fusion estimation based on the adaptive PF mainly includes three stages. First, several local filters perform in parallel to obtain individual sensor-based estimates. Second, all local estimates are combined in a master filter, where a global state estimate is yielded. Third, the global state and covariance is fed back to local filters with an information- sharing strategy for next cycle. The procedure of data fusion for the engine transient performance estimation is summarized as follows:

Step 1: Initialization

Given the initial values of global state x0, estimation error covariance P0, and process noise covariance Q0 in the master filter. The four local filters are initialized with the information allocation strategy:

$$Q_{j,0} = \beta_j^{-1} Q_{m,0} \quad P_{j,0} = \beta_j^{-1} P_{m,0} \quad X_{j,0} = \beta_j^{-1} X_{m,0} \quad j = 1, \cdots, 4 \quad (17)$$

where the information distribution factor βj follows:

$$\sum_{j=1}^{4} \beta_j = 1 \tag{18}$$

Step 2: The adaptive PF performs in the local filter.

The particles $\{x_{j,0:k-1}^i, w_{j,k-1}^i\}_{i=1}^{N}$ are generated based on the prior distribution $N(X_{j,k-1}, P_{j,k-1})$, and are propagated through the nonlinear model Equation (1). The numerical characteristics of the measured noise are computed by wavelet transformation, and the quantity σ is applied to the importance weight calculation. The known health information is imposed to the PF to produce the local constrained estimates of the state $\hat{x}_{j,k}$, error covariance $\hat{P}_{j,k}$ and noise covariance $Q_{j,k}$.

Step 3: Information fusion implements in the master filter.

The local estimates $\hat{x}_{j,k}$, $\hat{P}_{j,k}$, and $Q_{j,k}$ are sent to the master filter to fulfill the information fusion for the global optimal estimate xm,k. The estimate error covariance $\hat{P}_{j,k}$ is an important parameter representing the performance of the local filter, and it is used to calculate the fusing weight of the local filter.

The larger the covariance $\hat{P}_{j,k}$ is, the smaller the fusing weight in the global state is in the paper:

$$Q_{m,k} = (\sum_{j=1}^{4} Q_{j,k}^{-1})^{-1}$$

$$P_{m,k} = (\sum_{j=1}^{4} \hat{P}_{j,k}^{-1})^{-1} \qquad (19)$$

$$x_{m,k} = P_{m,k} \times \sum_{j=1}^{4} \hat{P}_{j,k}^{-1} x_{j,k}$$

Step 4: Information distribution strategy

The state estimate calculated by the master filter is transmitted back to each local filter with an information assignment strategy:

$$Q_{j,k} = \beta_j^{-1} Q_{m,k} \quad P_{j,k} = \beta_j^{-1} P_{m,k} \quad X_{j,k} = X_{m,k} \quad j = 1 \cdots N \qquad (20)$$

Due to the fact that information distribution factor β_j has no effect on estimation accuracy, it is set by $\beta_j = \dfrac{1}{N}$.

Steps (2)–(4) present the fusing PF algorithm at iteration k. For iteration $k + 1$, the state and covariance delay a time index and Steps (2)–(4) are repeated. The fusion filter architecture is shown in Figure 4 for gas turbine transient performance estimation.

As shown in the Figure , we can see that calculation loads are shared both by the local filters, and the master filter no longer undertakes the whole process like in the basic PF algorithm. Since the time update and measurement update are carried out independently in every local filter, the individual estimate by the local filter is not immediately affected by others. In the gas turbine engine health monitoring application, the data fusion filter architecture has a more efficient capability to deal with state estimation in cases of sensor fault due to fusing weight adaptive to estimation accuracy of the local filter.

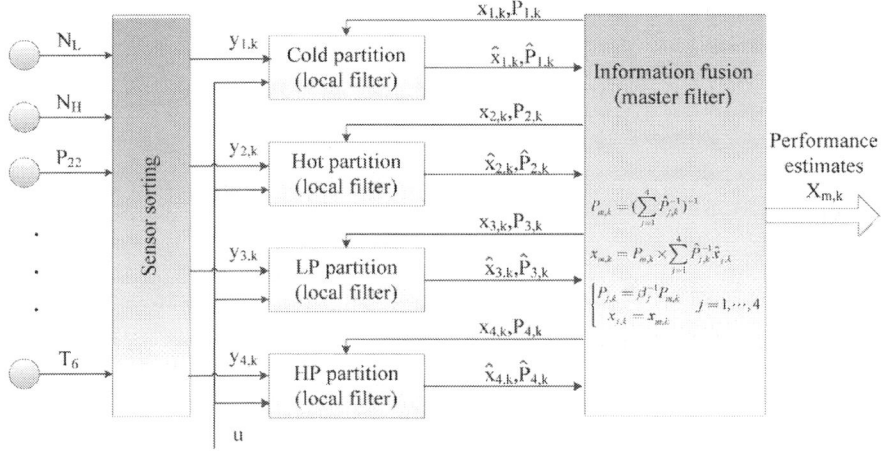

Figure 4. Data fusion filter architecture for gas turbine performance estimation.

4. SIMULATION AND ANALYSIS

The data fusion based on the adaptive PF approach is evaluated for the engine performance monitoring using the Matlab software. In the simulation environment, we use the component-level-model (CLM) engine to take the place of the actual engine, and the sampling rate equals to 50 Hz. The hardware of computer used for simulation is conFigure d as follows: CPU i3-2100 @ 3.10 GHz and RAM 2GB. The standard deviations of measurement and health parameter are shown in Table 1 andTable 2. Gaussian noise **v** with magnitude specified in Table 1 is added to the simulated measured values, and the independent system noise and initial measured noise separately follow $\omega \sim N(0,Q)$ and $v \sim N(0,R)$, wherein Q=0.16×10−4I6×6.

Table 1. Gas turbine component-level-model (CLM) model measurements, nominal value and standard deviation. High-pressure compressor: HPC; high-pressure turbine: HPT; low-pressure turbine: LPT.

Measurement	Acronyms	Normalized Value	Standard Deviation
Low pressure spool speed	N_L	1	0.0015
High pressure spool speed	N_H	1	0.0015
Fan outlet temperature	T_{22}	1	0.002
Fan outlet pressure	P_{22}	1	0.0015
HPC outlet temperature	T_3	1	0.002
HPC outlet pressure	P_3	1	0.0015
HPT outlet temperature	T_{43}	1	0.002
LPT outlet temperature	T_6	1	0.002

Gas turbine health condition is represented by health parameters as mentioned in the previous section. Table 2 shows four abrupt faults representative of possible situations expected to be encountered in practice, and the health parameter deviation in each case refers to lab record of the Rolls-Royce Company (London, UK) [3,28]. For example, Case 1 is a fan abrupt fault with its efficiency *SE1* deviating 1%. Assume that there are no fault on sensor measurements in the following experiments.

Table 2. Gas turbine engine abrupt fault modes and their deviation.

Scenarios	Acronyms	Fault Mode	Deviation	Standard Deviation
Case 1	SE1	Fan abrupt fault	−1% on SE1	0.0005
Case 2	SE2	HPC abrupt fault	−1% on SE2	0.0005
Case 3	SE3	HPT abrupt fault	−1% on SE3	0.0005
Case 4	SE4	LPT abrupt fault	−1% on SE4	0.0005

Engine gradual deterioration due to normal usage is simulated by linear drift of four health parameters, beginning from a healthy engine (the four parameters equal to 1) at cycle number $n = 0$ and with the degeneration at the end of the sequence at $n = 6000$: −2.18% on *SE1*, −6.71% on *SE2*, −3.22% on *SE3* and −0.81% on *SE4*. Considering the magnitude of both the engine abrupt fault and deterioration, the bounds of the health parameter representing the PF inequality constraints are separately set by $a = [1.005,1.005,1.005,1.005]^T$ and $b = [0.97,0.90,0.96,0.98]^T$. The performance of the engine anomaly detection is assessed by three indices, namely, root-mean-square error (*RMSE*), convergence time and root-mean-square deviation (*RMSD*). The *RMSE* and *RMSD* are separately defined by:

$$RMSE = [\frac{1}{S}\sum_{i=1}^{S}(\hat{x}_i - x_i)^2]^{\frac{1}{2}}$$

$$RMSD = [\frac{1}{S}\sum_{i=1}^{S}(\hat{x}_i - \bar{x}_i)^2]^{\frac{1}{2}}$$

(21)

where S is the sampling step and $\bar{x}_i$ the mean of estimate value. The convergence time T_c is used to indicate the delay in fault recognition. We define this time index in this paper that is from the starting deviation to the estimate steady state within ±0.02% range and no longer out of this range in two consecutive steps.

4.1. Abrupt Fault Diagnosis in Steady Operation Conditions

The tests on gas turbine abrupt fault diagnosis are first performed at ground steady conditions ($H = 0$ m, $Ma = 0$, $W_f = 2.48$ kg/s). The abrupt faults depicted in Table 2 are simulated, and the noise is not changed in the engine steady behavior. Given the stochastic character of the measurement noise, each test-case has been run five times. Then the engine fault diagnostic performances of basic KF, basic PF, fusion particle filter (F-PF) and fusion adaptive particle filter (FA-PF) are given in Table 3. The particle number of the PF is 60, and that of both the F-PF and FA-PF is 30.

The *RMSE*s and *RMSD*s of the two fusion PF approaches shown in Table 3 are smaller than those of the conventional PF and KF in the cases of the four abrupt fault modes, and the FA-PF one is superior to the others. The convergence time of the three PF approaches are nearly the same, and vary clearly in different cases. The KF is a linear estimator and it takes less time to reach the steady state. Two speed measurements are repeatedly utilized in each local filter and there is the reason for the fact the importance of their weights is increasing compared to the remaining measurements in the fusion filter structure. The fusion PF approaches have more satisfactory estimation accuracy due to sufficient extraction and information-sharing of key measurements such as the speeds for health monitoring. Because of the heuristic health knowledge enforced through inequality constraints, the FA-PF has the best estimation accuracy of fault diagnosis in the engine steady operation.

Table 3. The engine fault diagnostic performances of four filtering approaches in steady behavior. Kalman filter: KF; fusion particle filter: F-PF; fusion adaptive particle filter: FA-PF.

Fault Modes	Root-Mean-Square Error (RMSE)				Root-Mean-Square Deviation (RMSD)				T_c (ms)			
	KF	PF	F-PF	FA-PF	KF	PF	F-PF	FA-PF	KF	PF	F-PF	FA-PF
Case 1	0.0141	0.0108	0.0073	0.0059	0.0090	0.0085	0.0052	0.0044	220	190	230	220
Case 2	0.0137	0.0111	0.0078	0.0057	0.0094	0.0089	0.0059	0.0047	260	440	420	440
Case 3	0.0113	0.0118	0.0087	0.0060	0.0093	0.0096	0.0061	0.0050	320	620	660	600
Case 4	0.0126	0.0115	0.0080	0.0067	0.0100	0.0091	0.0058	0.0051	460	680	760	640

4.2. Abrupt Fault Diagnosis in Dynamic Operation

In order to further evaluate the proposed method performance in engine transient performance tracking, more simulation is carried out in the case of mixed gradual deterioration and abrupt faults. Gradual deterioration refers to all health parameters degrading linearly from the healthy condition to the end of 3000 cycles, and a simulated abrupt fault of magnitude −1% is added to *SE2* at 5 s. The engine operates in the dynamic behavior mode from $W_f =$

2.48 kg/s to 1.98 kg/s under three operation conditions: ground ($H = 0$ m, $Ma = 0$), high-altitude 1 ($H = 8000$ m, $Ma = 0.5$), and high-altitude 2 ($H = 11000$ m, $Ma = 0.8$). The particle number of the three PFs is set the same as previously, and the measurement noises remain unchanged. A comparison of the three approaches for the engine transient performance tracking in the cases of mixing gradual deterioration and abrupt fault at ground is depicted in Figure 5, where the dotted line and solid line are the real and estimated values of health parameters, respectively.

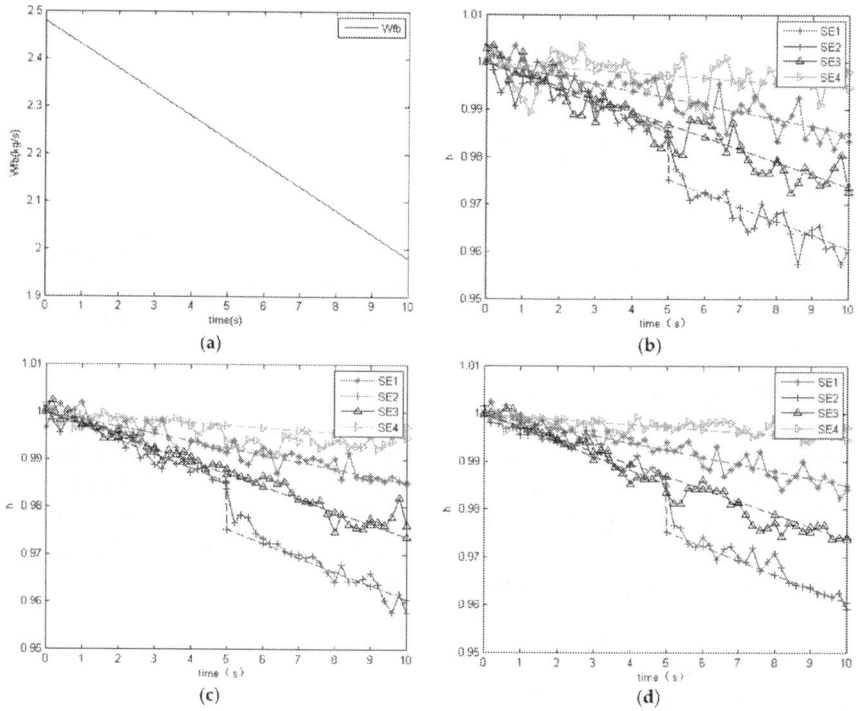

Figure 5. Engine transient performance tracking at ground condition. (a) Fuel supply rule W_f; (b) the performance estimates by the PF; (c) the performance estimates by the F-PF; and (d) the performance estimates by the FA-PF.

It can readily be found from the Figure 5 that the PF working in fusion architecture outperforms the basic PF. Table 4 further presents the estimation performance by the three PF methods in terms of number and data at different operating conditions. As can be seen from Table 4, the *RMSE* of the F-PF and FA-PF is smaller than that of PF, among which the FA-PF in three operation conditions are below 0.007. Hence, the estimation accuracy of fusion PF architecture outperforms that of the generic PF structure.

Table 4. The *RMSE* of dynamic estimation at three operation conditions.

Operation Condition	PF	F-PF	FA-PF
Ground	0.0120	0.0077	0.0069
High-altitude 1	0.0124	0.0076	0.0069
High-altitude 2	0.0123	0.0079	0.0070

4.3. Performance Estimation with Uncertain Noise in Dynamic Operation

The stochastic feature of the engine measured noise changes at different operation points. An experiment of gradual deterioration tracking with uncertain measurement noise is performed to assess the FA-PF algorithm. The engine experiences dynamic operation in the case of gradual deterioration as the same as the Section 4.2, but no abrupt fault is added. A series of the tests are implemented with ground conditions, including the change of only one measurement noise and all measurement noises simultaneously. Noise generated in the engine core, by sources such as the HPC, combustor, HPT and LPT, plays the most important roles to the overall noise under low-power conditions. While jet and fan noises have dominated over core noise at high engine power during takeoff [29].

The sensor noise varies with engine power condition in transient process. The uncertain noise of one sensor P_{22} in Figure 6 is simulated by the route of $R_0 =$ diag[0.0015, 0.0015, 0.002, 0.0015, 0.002, 0.0015, 0.002, 0.002] in 0–3 s, $R_1 =$ diag[0.0015, 0.0015, 0.002, 0.003, 0.002, 0.0015, 0.002, 0.002] in 3.02–6 s and $R_2 =$ diag[0.0015, 0.0015, 0.002, 0.0045, 0.002, 0.0015, 0.002, 0.002] in 6.02–10 s. The uncertain noise of each measurement undertakes the varied noise by R_0 in 0–3 s, $2R_0$ in 3.02–6 s, and $3R_0$ in 6.02–10 s in Figure 7. The size of the sliding window by wavelet transform is 50 steps.

As shown in Figure 6, the engine transient performance estimates by the fusion PF approaches seem to deviate from their real values once the noise of sensor P_{22} varies, no matter the quantity of measurement noise σ is equal to R_0, R_1 or R_2. The fusion adaptive PF has sound tracking performance due to the quantity σ adaptive to the real measurement noise real time in Figure 6d. We can obtain similar results in the case that all sensors for transient performance monitoring have uncertain measured noise in Figure 7.

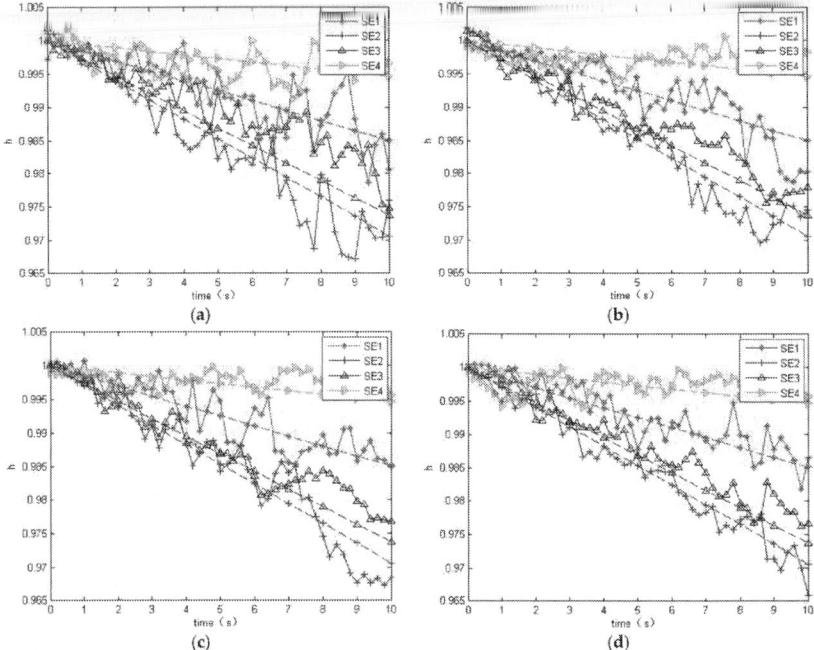

Figure 6. Transient performance monitoring with uncertain noise of one sensor P_{22} in the case of gradual deterioration. (**a**) σ = R_0; (**b**) σ = R_1; (**c**) σ = R_2; and (**d**) adaptive σ.

Table 5 summarizes the *RMSE* of the engine transient health estimation performance by the fusion PF approach with different the variable σ. The covariance of the measurement noise is estimated and real time tunes the quantity σ of the FA-PF algorithm as presented previously, and the *RMSE*s of the FA-PF are almost the same as those of the fusion PF where quantity σ is equal to the true value of measurement noise covariance in Table 5. Nevertheless, if the quantity σ is not tuned to the covariance of the real measurement noise, the fusion PF will produce a larger *RMSE* in terms of the engine transient performance estimation.

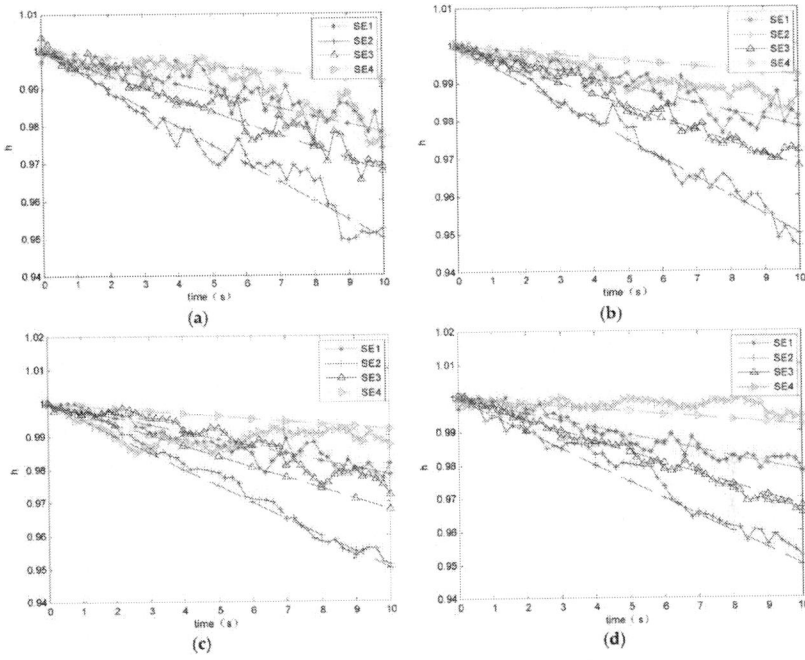

Figure 7. Transient performance monitoring with uncertain noise of every sensor in the case of gradual deterioration. (**a**) $\sigma = R_0$; (**b**) $\sigma = 2R_0$; (**c**) $\sigma = 3R_0$; and (**d**) adaptive σ.

Table 5. The *RMSE* of dynamic estimation with uncertain measurement noise at ground.

σ	Uncertain Noise of Sensor P_{22}				
	R_0	R_1	R_2	True Value	Tuning Value
RMSE	0.0121	0.0109	0.0108	0.0085	0.0088
σ	Uncertain Noise of All Sensors				
	R_0	$2R_0$	$3R_0$	True Value	Tuning Value
RMSE	0.0165	0.0116	0.0169	0.0095	0.0096

4.4. Engine Health Monitoring Test

Finally a test of the engine health monitoring is carried out to evaluate the proposed method at ground level. The engine input variables, W_f and A_8, are fed into the engine as shown in Figure 8. The engine *NH* representing the engine operation varies as follows: about 0.91 before 2.7 s, increasing from

0.91 to 1.0, then decreasing from 1.0 to 0.91, and about 0.91 to the end. During this process, the engine thrust increases from 0.819 to 1.0, and then back to 0.819. Four abrupt faults depicted in Table 2 are separately injected into the engine at 2 s.

Figure 9 depicts the variance estimates of measurement noise, which are applied to tune the quantity σ in the FA-PF method. The noise pollutes the true measurement, and it changes along the engine operation condition (the larger the power condition, the more noise enforcement). Table 6 summarizes the performance of engine health monitoring by four PF methods in the abrupt fault cases. The fusion PF with inequality constraints is defined by the FC-PF. The detection results in Table 6 show that the FC-PF and FA-PF have the less estimation errors and provide more stable estimates than the other PF methods because of the prior knowledge used. Furthermore, the performance of the FA-PF for engine health monitoring is the best due to the quantity of the importance function σ adaptive to the measurement noise real-time.

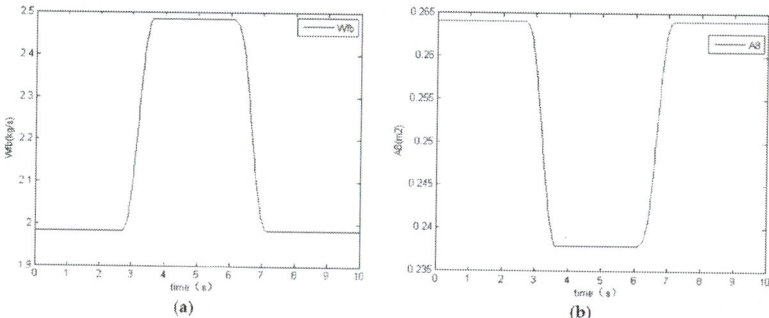

Figure 8. The change rules of the engine input variables. (**a**) W_f; and (**b**) A_8.

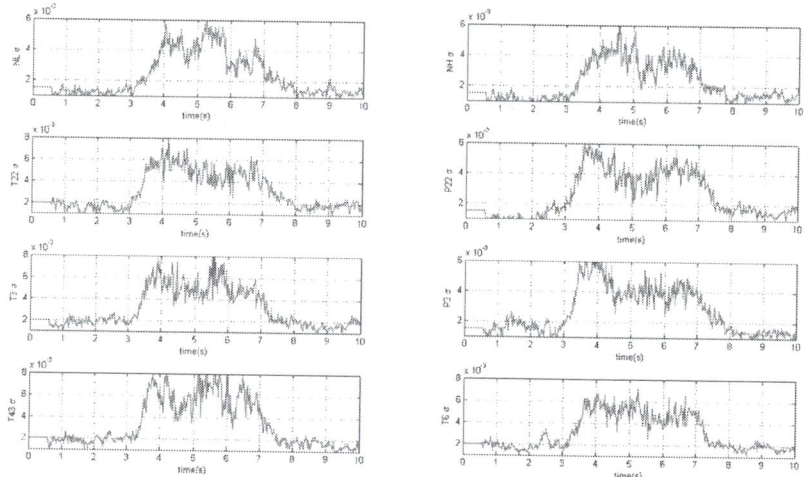

Figure 9. The variance estimates of measurement noise in the FA-PF.

Table 6. Engine health monitoring performance by four PF methods in the cases of abrupt fault.

Fault Modes	RMSE				RMSD			
	PF	F-PF	FC-PF	FA-PF	PF	F-PF	FC-PF	FA-PF
Case 1	0.0140	0.0101	0.0084	0.0061	0.0123	0.0083	0.0075	0.0050
Case 2	0.0138	0.0095	0.0078	0.0055	0.0119	0.0079	0.0069	0.0043
Case 3	0.0134	0.0100	0.0089	0.0058	0.0117	0.0069	0.0079	0.0050
Case 4	0.0144	0.0106	0.0094	0.0065	0.0132	0.0077	0.0079	0.0052

5. CONCLUSIONS

This paper describes the use of data fusion based on the adaptive PF for gas turbine dynamic performance monitoring. The state estimation in the fusion estimator architecture includes three steps: several local filters working in parallel to obtain individual sensor-based estimates, one master filter fusing these local estimates to yield a global state estimate, and the global estimates serving as the feedback to each local filter with information-sharing strategy. A systematic comparison of the fusion PF methods for transient performance estimation is presented. Gradual deterioration, abrupt faults and their mixtures are typically considered as the engine anomaly scenarios in the test. The fusion PF architecture has better estimation accuracy and less convergence time than the conventional PF architecture. The implementation of the fusion PF is quite straightforward and involves only basic matrix operations. The convergence time by the fusion PF is similar to that by the basic PF, yet the computational burden of the master filter is reduced, because it is shared by local filters in the fusion PF structure.

Moreover, an adaptive PF algorithm is proposed to sufficiently utilize prior information for the engine transient performance detection with measurement noise uncertainty. The heuristic health knowledge is usually neglected in model-based engine diagnoses due to the complex mathematics application to the conventional PF. In this paper, the engine prior health information represented by the inequality constraints is enforced in the fusion PF algorithm. In addition, the uncertainty of the measurement noise in the engine dynamic operation is considered in the fusion PF. The covariance of the sensed noise is estimated and then applied to tune the importance weights of the PF. The improvements brought by the data fusion based on the adaptive PF have been illustrated on the application of the engine transient performance monitoring with noise uncertainty. The experiments show that the proposed method leads to the more reliable assessments of the engine health conditions no matter whether the cases involve gradual engine deterioration or abrupt faults. With information fusion, *a priori* knowledge and measurement noise adaption in mind, the data fusion based on the adaptive PF approach seems therefore to be more promising. The present work by the authors has shown the advantages of data fusion based on an adaptive PF for gas turbine health monitoring. It is not considered that the phase errors between different measurements combined with the senor lag will affect the shift in engine performance, and it would be interesting to discuss the proposed approach of this paper with addition of measurement differences.

ACKNOWLEDGMENTS

We are grateful for the financial support of the National Nature Science Foundation of China (No. 61304133), the Fundamental Research Funds for the Central Universities (No. NS2015024). Moreover, the authors wish to thank the anonymous reviewers for their constructive comments and great help in the writing process, which improve the manuscript significantly.

AUTHOR CONTRIBUTIONS

Feng Lu and Jinquan Huang contributed in developing the ideas of this research, Yafan Wang and Yihuan Huang performed this research. All of the authors were involved in preparing this manuscript.

REFERENCES

1. Volponi, A. Gas turbine engine health management: Past, present, and future trends. *J. Eng. Gas Turbines Power* **2014,***136*.
2. Rodger, J.A. Toward reducing failure risk in an integrated vehicle health maintenance system: A fuzzy multi-sensor data fusion Kalman filter approach for IVHMS. *Expert Syst. Appl.* **2012**, *39*, 9821–9836.
3. Borguet, S.; Léonard, O. Comparison of adaptive filters for gas turbine performance monitoring. *J. Comput. Appl. Math.* **2010**, *234*, 2202–2212.
4. Li, Y.G.; Korakianitis, T. Nonlinear weighted-least-squares estimation approach for gas-turbine diagnostic applications. *J. Propuls. Power* **2011**, *27*, 337–345.
5. Joly, R.B.; Ogaji, S.O.T.; Singh, R.; Probert, S.D. Gas-turbine diagnostics using artificial neural-networks for a high bypass ratio military turbofan engine. *Appl. Energy* **2004**, *78*, 397–418.
6. Vanini, Z.N.S.; Khorasani, K.; Meskin, N. Fault detection and isolation of a dual spool gas turbine engine using dynamic neural networks and multiple model approach. *Inf. Sci.* **2014**, *259*, 234–251.
7. Eustace, R.W. A real-world application of fuzzy logic and influence coefficients for gas turbine performance diagnostics. *J. Eng. Gas Turbines Power* **2008**, *130*.

8. Simon, D. A comparison of filtering approaches for aircraft engine health estimation. *Aerosp. Sci. Technol.* **2008**, *12*, 276–284.

9. Sun, J.G.; Vasilyev, V.; Ilyasov, B. *Advanced Multivariable Control Systems of Aeroengines*; Beihang Press: Beijing, China, 2005; pp. 60–83.

10. Lu, F.; Chen, Y.; Huang, J.Q.; Zhang, D.D. An integrated nonlinear model-based approach to gas turbine engine sensor fault diagnostics. *J. Aerosp. Eng.* **2014**, *228*, 2007–2021.

11. Volponi, A. *Enhanced Self Tuning On-Board Real-Time Model (eSTORM) for Aircraft Engine Performance Health Tracking*; Technical Report for National Aeronautics and Space Administration: Cleveland, OH, USA, 2008.

12. Armstrong, J.B.; Simon, D.L. *Implementation of an Integrated On-Board Aircraft Engine Diagnostic Architecture*; Technical Report for National Aeronautics and Space Administration: Cleveland, OH, USA, 2012.

13. Simon, D. Kalman filtering with state constraints: A survey of linear and nonlinear algorithms. *IET Control Theory Appl.* **2010**, *4*, 1303–1318.

14. Lu, F.; Huang, J.Q.; Lv, Y.Q. Gas path health monitoring for a turbofan engine based on a nonlinear filtering approach. *Energies* **2013**, *6*, 492–513.

15. Gordon, N.; Salmond, D.; Smith, A. Novel Approach to Nonlinear/Non-Gaussian Bayesian State Estimation. *IEE Proc. F Radar Signal Process.* **1993**, *140*, 107–113.

16. Climente-Alarcon, V.; Antonino-Daviu, J.A.; Haavisto, A.; Arkkio, A. Particle filter-based estimation of instantaneous frequency for the diagnosis of electrical asymmetries in induction machines. *IEEE Trans. Instrum. Meas.* **2014**, *63*, 2454–2463.

17. Zhao, B.; Skjetne, R.; Blanke, M.; Dukan, F. Particle filter for fault diagnosis and robust navigation of underwater robot. *IEEE Trans. Control Syst. Technol.* **2014**, *22*, 2399–2407.

18. Tao, G.L.; Deng, Z.L. Self-tuning fusion Kalman filter for multisensor single-channel ARMA signals with coloured noises. *IMA J. Math. Control Inf.* **2015**, *32*, 55–74.

19. Zhu, H.Y.; Zhai, Q.Z.; Yu, M.W.; Han, C.Z. Estimation fusion algorithms in the presence of partially known cross-correlation of local estimation errors. *Inf. Fusion* **2014**, *18*, 187–196.

20. Seifzadeh, S.; Khaleghi, B.; Karray, F. Distributed soft-data-constrained multi-model particle filter. *IEEE Trans. Cybern.* **2015**, *45*, 384–394.

21. Zajac, M. Online fault detection of a mobile robot with a parallelized particle filter. *Neurocomputing* **2014**, *126*, 151–165.

22. Li, T.C.; Sun, S.D.; Sattar, T.P. Fight sample degeneracy and impoverishment in particle filters: A review of intelligent approaches. *Expert Syst. Appl.* **2014**, *41*, 3944–3954.

23. Simon, D. *Constrained Kalman Filtering via Density Function Truncation for Turbofan Engine Health Estimation*; Technical Report for National Aeronautics and Space Administration: Cleveland, OH, USA, 2006.

24. Boulkroune, B.; Darouach, M.; Zasadzinski, M. Moving horizon state estimation for linear discrete-time singular systems. *IET Control Theory Appl.* **2010**, *4*, 339–350.

25. Yadav, S.K.; Sinha, R.; Bora, P.K. Electrocardiogram signal denoising using non-local wavelet transform domain filtering. *IET Signal Process.* **2015**, *9*, 88–96.

26. Kyriazis, A.; Mathioudakis, K. Enhance of fault localization using probabilistic fusion with gas path analysis algorithms. *J. Eng. Gas Turbines Power* **2009**, *131*.

27. Kaltungo, A.Y.; Sinha, J.K.; Elbhbah, K. An improved data fusion technique for faults diagnosis in rotating machines. *Measurement* **2014**, *58*, 27–32.

28. Curnock, B. *Obidicote Project—Work Package 4: Steady-State Test Cases*; Rolls-Royce PLC: Manchester, UK, 2000.

29. Hultgren, L.S.; Miles, J.H. Noise-Source Separation Using Internal and Far-Field Sensors for a Full-Scale Turbofan Engine. In Proceedings of the 15th AIAA/CEAS Aeroacoustics Conference (30th AIAA Aeroacoustics Conference) Cosponsored by AIAA and CEAS, Miami, FL, USA, 11–13 May 2009.

Chapter 4

Robust Hammerstein Adaptive Filtering under Maximum Correntropy Criterion

Zongze Wu [1], **Siyuan Peng** [1], **Badong Chen** [2,*] **and Haiquan Zhao** [3]

[1] *School of Electronic and Information Engineering, South China University of Technology, Guangzhou 510640, China*
[2] *School of Electronic and Information Engineering, Xi'an Jiaotong University, Xi'an 710049, China*
[3] *School of Electrical Engineering, Southwest Jiaotong University, Chengdu 610031, China*

ABSTRACT

The maximum correntropy criterion (MCC) has recently been successfully applied to adaptive filtering. Adaptive algorithms under MCC show strong robustness against large outliers. In this work, we apply the MCC criterion to develop a robust Hammerstein adaptive filter. Compared with the traditional Hammerstein adaptive filters, which are usually derived based on the well-known mean square error (MSE) criterion, the proposed algorithm can achieve better convergence performance especially in the presence of impulsive non-Gaussian (e.g., α-stable) noises. Additionally, some theoretical results concerning the convergence behavior are also obtained. Simulation examples are presented to confirm the superior performance of the new algorithm.

Keywords: Hammerstein adaptive filtering; MCC; nonlinear system identification

1. INTRODUCTION

Nonlinear system identification is still an active research area [1]. Although linear systems have established a solid theory [2], most practical systems (e.g., hands-free telephone systems) may be more adequately represented as a nonlinear model. One of the main challenges for nonlinear system identification is the choice of an appropriate nonlinear filtering structure that accurately captures the characteristics of the underlying nonlinear system. A common structure used in nonlinear modeling is the *block-oriented* representation. The Wiener model and the Hammerstein model are two typical block-oriented nonlinear models [3]. Specifically, the Wiener model consists of a cascade of a linear time invariant (LTI) filter followed by a static nonlinear function, indicated as a linear-nonlinear (LN) model [4,5,6], and the Hammerstein model consists of a cascade of a static nonlinear function follow by a LTI filter, known as a nonlinear-linear (NL) model [7,8,9,10,11,12,13,14,15,16,17,18,19]. Other nonlinear models include neural networks (NNs) [20], Volterra adaptive filters (VAFs) [21], kernel adaptive filters (KAF) [22,23,24,25], among others.

Hammerstein filters can accurately model many real-world systems and, as a consequence, they have been successfully used in various applications of engineering [26,27,28,29]. Due to its simplicity and efficiency, the mean square error (MSE) criterion has been widely applied in Hammerstein adaptive filtering [30]. Adaptive algorithms under MSE usually perform very well when the desired signals are disturbed by Gaussian noises. However, when the desired signals are disturbed by non-Gaussian noises, especially in the presence of large outliers (observations that significantly deviate from the bulk of data), the performance of the MSE based algorithms may deteriorate rapidly. Actually, MSE is rather sensitive to outliers. In most practical situations, heavy-tailed impulsive noises may occur, which often cause large outliers. For instance, different types of artificial noises in electronic devices, atmospheric noises, and lighting spikes in natural phenomena, can be described as an impulsive noise [31,32].

In this work, instead of using the MSE criterion, we apply the maximum correntropy criterion (MCC) to develop a robust Hammerstein adaptive filtering algorithm. Correntropy is a nonlinear similarity measure between two signals [33,34]. The MCC aims at maximizing the similarity (measured by correntropy) between the model output and the desired response such that the adaptive model is as close as possible to the unknown system. It has been shown that, the MCC in terms of the stability and accuracy, is very robust with respect to impulsive noises [33,34,35,36,37,38,39]. Compared with the traditional Hammerstein adaptive filtering algorithms based on the MSE criterion, the new algorithm can achieve better performance especially in the presence of impulsive non-Gaussian noises.

The organization of the rest of the paper is as follows. In Section 2, after briefly introducing the correntropy, we derive a Hammerstein adaptive filtering algorithm under MCC criterion. In Section 3, we carry out the convergence analysis. InSection 4, we present simulation examples to demonstrate the superior performance of the proposed algorithm. Finally, we give the conclusion in Section 5.

2. HAMMERSTEIN ADAPTIVE FILTERING UNDER THE MAXIMUM CORRENTROPY CRITERION

Figure 1 shows the structure of a Hammerstein adaptive filter under MCC criterion, where the filter consists of a polynomial memoryless nonlinearity followed by a linear FIR filter. This structure has been commonly used in Hammerstein adaptive filtering [8,9,27]. As shown in Figure 1, under the MCC criterion, the parameters of the linear and nonlinear parts are adjusted to maximize the correntropy between the model output and desired response.

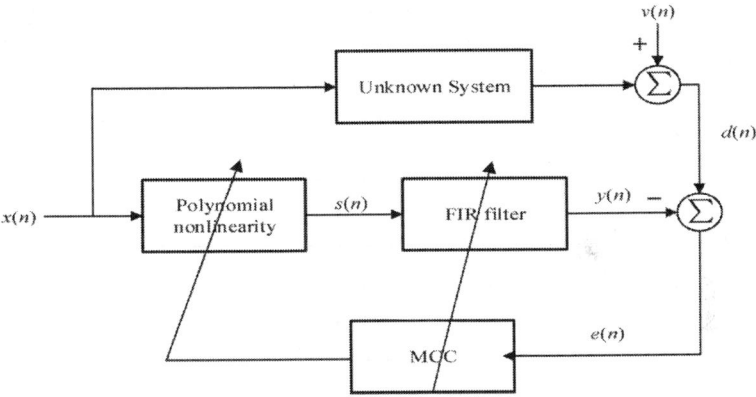

Figure 1. Structure of a Hammerstein adaptive filter under maximum correntropy criterion (MCC) criterion.

2.1. Correntropy

Correntropy is a nonlinear similarity measure between two signals. Given two random variables X and Y, the correntropy is [33,34,35,36,37,38,39]

$$V(X,Y) = E[\kappa(X,Y)] = \int \kappa(x,y) f_{XY}(x,y) dxdy \qquad (1)$$

where $E[\cdot]$ denotes the expectation operator, $\kappa(\cdot,\cdot)$ is a *shift-invariant* Mercer kernel, and $f_{XY}(x, y)$ stands for the probability density function (PDF) of (X, Y). The most widely used kernel in correntropy is the Gaussian kernel, given by

$$\kappa_\sigma(x, y) = \frac{1}{\sigma\sqrt{2\pi}} \exp(-\frac{e^2}{2\sigma^2}) \tag{2}$$

where $e = x - y$, and σ stands for the kernel bandwidth. In this work, without being mentioned otherwise, the kernel function is a Gaussian kernel. In practical situations, the join distribution of X and Y is usually unknown and only a finite number of data $\{(d(i), y(i))\}_{i=1}^{K}$ are available. In these cases, one can use a sample mean estimator of the correntropy:

$$\hat{V}_{N,\sigma}(X,Y) = \frac{1}{K}\sum_{i=1}^{K}\kappa_\sigma(d(i) - y(i)) \tag{3}$$

The optimization cost under MCC is thus

$$\max J_{MCC} = \sum_{i=1}^{K}\kappa_\sigma(e(i)) \tag{4}$$

where $e(i) = d(i) - y(i)$. We can evaluate the sensitivity (derivative) of the MCC cost J_{MCC} with respect to the error $e(i)$,

$$\frac{\partial J_{MCC}}{\partial e(i)} = \frac{-1}{\sqrt{2\pi}\sigma^3} \cdot \exp\left(\frac{-e^2(i)}{2\sigma^2}\right)e(i) \tag{5}$$

The derivative curves of $-J_{MCC}$ for different kernel widths are illustrated in Figure 2. As one can see, when the magnitude of error is very large, the derivative will become rather small especially for a smaller kernel width. Therefore, the MCC training is insensitive (hence robust) to a large error.

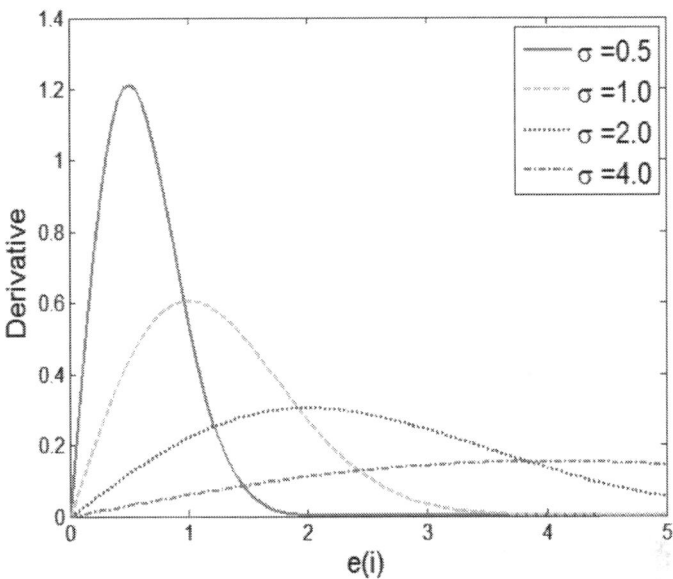

Figure 2. Derivative curves of $-J_{MCC}$ with respect to $e(i)$ for different kernel widths.

2.2. Hammerstein Adaptive Filtering

Assuming that the input-output mapping of the memoryless polynomial nonlinearity is

$$s(n) = p_1 x(n) + p_2 x^2(n) + \cdots + p_M x^M(n) \tag{6}$$

where M and p_M denote the polynomial order and the m-th order coefficient, Expression (6) can be rewritten as

$$s(n) = \mathbf{p}^T(n)\mathbf{x}_p(n) \tag{7}$$

where $\mathbf{x}_p(n) = [x(n)\, x^2(n)\cdots x^M(n)]^T$ is the polynomial regressor, and $\mathbf{p}(n) = [p_1\, p_2\cdots p_M]^T$ is the polynomial coefficient vector. The output of the FIR filter can be expressed as

$$y(n) = \mathbf{w}^T(n)\mathbf{s}(n) \tag{8}$$

where $\mathbf{w}(n) = [w_0\, w_1\cdots w_{N-1}]^T$ is the FIR weight vector, and $\mathbf{s}(n) = [s(n)\, s(n - 1)\cdots s(n - N + 1)]^T$ is the FIR input vector, with N being the FIR memory size. Let $X(n) = [\mathbf{x}_p(n)\, \mathbf{x}_p(n - 1)\cdots \mathbf{x}_p(n - N + 1)]^T$. Then we have

$$s(n) = X^T(n)p(n) \tag{9}$$

Combining Equations (8) and (9) yields

$$y(n) = w^T(n)s(n) = w^T(n)X^T(n)p(n) \tag{10}$$

Assume that the unknown system that needs to be identified is also a Hammerstein system with parameter vectors $p* = [p*1\ p*2\cdots p*M]T$ and $w* = [w*0\ w*1\cdots w*N-1]T$. Then, the desired signal can be expressed as

$$d(n) = w^{*T}X^T(n)p^* + v(n) \tag{11}$$

where $v(n)$ stands for an additive disturbance noise. The error signal can then be calculated as $e(n) = d(n) - w^T(n)X^T(n)p(n)$. In the following, we derive an adaptive algorithm to estimate the Hammerstein parameter vectors using MCC instead of MSE as an optimization criterion. Let us consider the following cost function

$$J_{MCC}(\mathbf{p},\mathbf{w}) = \sum_{j=n-L+1}^{n} \kappa_\sigma(d(j), y(j))$$

$$= \frac{1}{\sqrt{2\pi}\sigma} \sum_{j=n-L+1}^{n} \exp\left(\frac{-e^2(j)}{2\sigma^2}\right) \tag{12}$$

where $e(j) = d(j) - y(j)$, and L denotes the sliding data length. Then, a steepest ascent algorithm for estimating the polynomial coefficient vector can be derived as follows:

$$\frac{\partial J_{MCC}(\mathbf{p},\mathbf{w})}{\partial \mathbf{p}(n)} = \frac{1}{\sqrt{2\pi}\sigma^3} \cdot \sum_{j=n-L+1}^{n} \exp\left(\frac{-e^2(j)}{2\sigma^2}\right) e(j) \cdot \frac{\partial e(j)}{\partial \mathbf{p}(j)}$$

$$= \frac{1}{\sqrt{2\pi}\sigma^3} \cdot \sum_{j=n-L+1}^{n} \exp\left(\frac{-e^2(j)}{2\sigma^2}\right) e(j) \cdot X(j)\mathbf{w}(j) \tag{13}$$

$$\mathbf{p}(n+1) = \mathbf{p}(n) + \mu_p \frac{\partial J_{MCC}(\mathbf{p},\mathbf{w})}{\partial \mathbf{p}(n)} = \mathbf{p}(n) + \frac{\mu_p}{\sqrt{2\pi}\sigma^3} \cdot \sum_{j=n-L+1}^{n} \exp\left(\frac{-e^2(j)}{2\sigma^2}\right) e(j) \cdot X(j)\mathbf{w}(j) \tag{14}$$

In a similar way, we propose the following weight update equation for the coefficients of the FIR filter:

$$\frac{\partial J_{MCC}(\mathbf{p},\mathbf{w})}{\partial \mathbf{w}(n)} = \frac{1}{\sqrt{2\pi\sigma^3}} \cdot \sum_{j=n-L+1}^{n} \exp\left(\frac{-e^2(j)}{2\sigma^2}\right) e(j) \cdot \frac{\partial e(j)}{\partial \mathbf{w}(j)}$$

(15)

$$= \frac{1}{\sqrt{2\pi\sigma^3}} \cdot \sum_{j=n-L+1}^{n} \exp\left(\frac{-e^2(j)}{2\sigma^2}\right) e(j) \cdot \mathbf{X}^T(j)\mathbf{p}(j)$$

$$\mathbf{w}(n+1) = \mathbf{w}(n) + \mu_w \frac{\partial J_{MCC}(\mathbf{p},\mathbf{w})}{\partial \mathbf{w}(n)} = \mathbf{w}(n) + \frac{\mu_w}{\sqrt{2\pi\sigma^3}} \cdot \sum_{j=n-L+1}^{n} \exp\left(\frac{-e^2(j)}{2\sigma^2}\right) e(j) \cdot \mathbf{X}^T(j)\mathbf{p}(j)$$

(16)

In Equations (14) and (16), μ_p and μ_w are, respectively, step-sizes for polynomial nonlinearity subsystem and FIR subsystem. In this work, for simplicity we consider only the stochastic gradient based algorithm (*i.e.*, $L = 1$). In this case, we have

$$\mathbf{p}(n+1) = \mathbf{p}(n) + \eta_p \cdot \exp\left(\frac{-e^2(n)}{2\sigma^2}\right) e(n) \cdot \mathbf{X}(n)\mathbf{w}(n)$$

(17)

$$\mathbf{w}(n+1) = \mathbf{w}(n) + \eta_w \cdot \exp\left(\frac{-e^2(n)}{2\sigma^2}\right) e(n) \cdot \mathbf{X}^T(n)\mathbf{p}(n)$$

(18)

where $\eta_p = \frac{\mu_p}{\sqrt{2\pi\sigma^3}}$, and $\eta_w = \frac{\mu_w}{\sqrt{2\pi\sigma^3}}$. The above update equations are referred to as the Hammerstein adaptive filtering algorithm under MCC criterion, whose pseudocodes are presented in Algorithm 1. The proposed algorithm is in form similar to the traditional Hammerstein adaptive filters under MSE criterion [7], but the step-sizes are different.

Algorithm 1: Hammerstein adaptive filtering Algorithm under MCC.

Parameters setting: μ_p, μ_w, σ

Initialization: $\mathbf{p}(0)$, $\mathbf{w}(0)$

For $n = 1, 2, \ldots$ **do**

   (1) $s(n) = \mathbf{X}^T(n)\mathbf{p}(n)$

   (2) $y(n) = \mathbf{w}^T(n)s(n)$

   (3) $e(n) = d(n) - y(n)$

   (4) $\mathbf{p}(n+1) = \mathbf{p}(n) + \eta_p \cdot \exp\left(\frac{-e^2(n)}{2\sigma^2}\right) e(n) \cdot \mathbf{X}(n)\mathbf{w}(n)$

   (5) $\mathbf{w}(n+1) = \mathbf{w}(n) + \eta_w \cdot \exp\left(\frac{-e^2(n)}{2\sigma^2}\right) e(n) \cdot \mathbf{X}^T(n)\mathbf{p}(n)$

End for

---

3. CONVERGENCE ANALYSIS

3.1. Stability Analysis

Using the Taylor series expansion of the error $e(n + 1)$ around the instant n and keeping only the linear term, we have [4,7,40]

$$e(n+1) = e(n) + \frac{\partial e(n)}{\partial p(n)}\bigg|_{w(n)=const} \Delta p(n) + \frac{\partial e(n)}{\partial w(n)}\bigg|_{p(n)=const} \Delta w(n) + h.o.t$$

(19)

where $h.o.t$ denotes higher-order terms. Combining Equations (11), (17) and (18), we can obtain

$$\frac{\partial e(n)}{\partial p(n)} = -X(n)w(n)$$

(20)

$$\frac{\partial e(n)}{\partial w(n)} = -X^T(n)p(n)$$

(21)

$$\Delta p(n) = \eta_p \cdot \exp\left(\frac{-e^2(n)}{2\sigma^2}\right) e(n) \cdot X(n)w(n)$$

(22)

$$\Delta w(n) = \eta_w \cdot \exp\left(\frac{-e^2(n)}{2\sigma^2}\right) e(n) \cdot X^T(n)p(n)$$

(23)

Substituting Equations (20)–(23) in Equation (19), and after simple manipulation, we have

$$e(n+1) = \left[1 - \eta_p \cdot \exp\left(\frac{-e^2(n)}{2\sigma^2}\right) \cdot \|X(n)w(n)\|^2 - \eta_w \cdot \exp\left(\frac{-e^2(n)}{2\sigma^2}\right) \cdot \|X^T(n)p(n)\|^2\right] e(n)$$

(24)

To ensure the stability of the proposed algorithm, we must assure that $|e(n + 1)| \leq |e(n)|$, and hence

$$\left| 1 - \eta_p \cdot \exp\left(\frac{-e^2(n)}{2\sigma^2} \right) \cdot \|X(n)w(n)\|^2 - \eta_w \cdot \exp\left(\frac{-e^2(n)}{2\sigma^2} \right) \cdot \|X^T(n)p(n)\|^2 \right| \leq 1 \tag{25}$$

which yields

$$0 < \eta_p \cdot \|X(n)w(n)\|^2 + \eta_w \cdot \|X^T(n)p(n)\|^2 \leq \frac{2}{\exp\left(\dfrac{-e^2(n)}{2\sigma^2} \right)} \tag{26}$$

Since $\left(\dfrac{-e^2(n)}{2\sigma^2} \right) \leq 1$, the following condition guarantees convergence:

$$0 < \eta_p \cdot \|X(n)w(n)\|^2 + \eta_w \cdot \|X^T(n)p(n)\|^2 \leq 2 \tag{27}$$

Remark 1. The derived bound on step-sizes is only of theoretical importance as in general, Equation (27) cannot be verified in a practical situation. Similar theoretical results can be found in [7].

3.2. Steady-State Mean Square Performance

We denote $e_{pw}(n)$ the *a priori* error of the whole system, $e_p(n)$ the *a priori* error when only the nonlinear part is adapted while the linear filter is fixed, and $e_w(n)$ the *a priori* error when only the linear filter is adapted while the nonlinear part is fixed. Let $H_p = \lim_{n \to \infty} E\left[e_p^2(n) \right]$, $H_w = \lim_{n \to \infty} E\left[e_w^2(n) \right]$, and $H_{pw} = \lim_{n \to \infty} E\left[e_{pw}^2(n) \right]$ be the steady-state *excess* *mean* *square* errors (EMSEs). In addition, we denote

$$f(e(i)) = \exp\left(\frac{-e^2(i)}{2\sigma^2} \right) e(i) \tag{28}$$

Before evaluating the theoretical values of the steady-state EMSEs, we make the following assumptions:

(A) The noise $v(n)$ is zero-mean, independent, identically distributed, and is independent of the input X(n), $\hat{s}(n)$ and $e(n)$.

(B) The *a priori* errors $e_p(n)$ and $e_w(n)$ are zero-mean Gaussian, and independent of the noise $v(n)$.

(C) $\|X(n)w(n)\|^2$ and $\|X^T(n)p(n)\|^2$ are   asymptotically   uncorrelated with $f^2(e(n))$, that is

$$\lim_{n\to\infty} E\left[\|X(n)w(n)\|^2 f^2(e(n))\right] = Tr(R_{WX})\lim_{n\to\infty} E\left[f^2(e(n))\right] \tag{29}$$

$$\lim_{n\to\infty} E\left[\|X^T(n)p(n)\|^2 f^2(e(n))\right] = Tr(R_{XP})\lim_{n\to\infty} E\left[f^2(e(n))\right] \tag{30}$$

where $R_{WX} = E\left[(X(n)w(n))(X(n)w(n))^T\right]$ and $R_{XP} = E\left[(X^T(n)p(n))(X^T(n)p(n))^T\right]$ are the covariance matrices, and Tr(·) denotes the trace operator.

Remark 2. For the assumption (A), it is very common to assume that the noise is independent of the regression vector [41,42,43]. In addition, the noise is often restricted to be zero-mean, identically distributed [33,34,35]. As discussed in [44,45], the assumption (B) is reasonable for long adaptive filters. Since $e_p(n)$ is the *a priori error* when only the nonlinear part is adapted while the linear filter is fixed, we have the approximation $w^* \approx w(n)$ such that $w(n)$ is asymptotically uncorrelated with $f^2(e(n))$. Due to the independent assumption (A), $X(n)$ is also asymptotically uncorrelated with $f^2(e(n))$. So $\|X^T(n)w(n)\|^2$ is asymptotically uncorrelated with $f^2(e(n))$. Similarly, $\|X^T(n)p(n)\|^2$ is asymptotically uncorrelated with $f^2(e(n))$. Therefore, the assumption (C) is rational.

When only the polynomial part with parameter vector **p** is adapted, the error $e_p(n)$ is

$$e_p(n) = w^{*T}X^T(n)p^* - w^T(n)X^T(n)p(n) \approx w^T(n)X^T(n)\tilde{p}(n) \tag{31}$$

where $\tilde{p}(n) = p^* - p^T(n)$.   In   Equation   (31),   we   use   the approximation $w^* \approx w(n)$ at steady-state. From Equation (17), it follows easily that

$$\tilde{p}(n+1) = \tilde{p}(n) - \eta_p \cdot f(e(n)) \cdot X(n)w(n) \tag{32}$$

Squaring both sides of Equation (32), we have

$$\|\tilde{p}(n+1)\|^2 = \|\tilde{p}(n)\|^2 - 2\eta_p \cdot f(e(n)) \cdot e_p(n) + \eta_p^2 \cdot f^2(e(n)) \cdot \|X(n)w(n)\|^2 \tag{33}$$

Taking the expectations of the both sides of Equation (33) yields

$$E\left[\|\tilde{p}(n+1)\|^2\right] = E\left[\|\tilde{p}(n)\|^2\right] - 2\eta_p \cdot E\left[f(e(n)) \cdot e_p(n)\right] + \eta_p^2 \cdot E\left[f^2(e(n)) \cdot \|X(n)w(n)\|^2\right] \tag{34}$$

Assuming the filter is stable and attains the steady state, it holds

$$\lim_{n\to\infty} E\left[\|\tilde{p}(n+1)\|^2\right] = \lim_{n\to\infty} E\left[\|\tilde{p}(n)\|^2\right] \tag{35}$$

Combining Equations (34) and (35) and the above assumptions, we obtain

$$2 \cdot \lim_{n \to \infty} E\left[f(e(n)) \cdot e_p(n) \right] = \eta_p Tr(R_{WX}) \lim_{n \to \infty} E\left[f^2(e(n)) \right] \quad (36)$$

In order to derive a theoretical value of the steady-state EMSE, we consider two cases below.

Case A. Gaussian Noise

Recalling that $e(n) = e_p(n) + v(n)$, and assuming that the noise $v(n)$ is zero-mean Gaussian, with variance ς_v^2, we get [34]

$$\lim_{n \to \infty} E\left[f(e(n)) \cdot e_p(n) \right] = \frac{\sigma^3 H_p}{(\sigma^2 + \varsigma_v^2 + H_p)^{3/2}} \quad (37)$$

where σ_e^2 denotes the variance of the error, and $\sigma_e^2 = E[e_p^2(n)] + \varsigma_v^2$. Similarly, we obtain [29]

$$\lim_{n \to \infty} E\left[f^2(e(n)) \right] = \frac{\sigma^3 (H_p + \varsigma_v^2)}{(\sigma^2 + 2\varsigma_v^2 + 2H_p)^{3/2}} \quad (38)$$

Substituting Equations (37) and (38) into Equation (36), we have

$$2 \cdot \frac{\sigma^3 H_p}{(\sigma^2 + \varsigma_v^2 + H_p)^{3/2}} = \eta_p Tr(R_{WX}) \frac{\sigma^3 (H_p + \varsigma_v^2)}{(\sigma^2 + 2\varsigma_v^2 + 2H_p)^{3/2}} \quad (39)$$

Therefore, the steady-state EMSE H_p satisfies

$$H_p = \frac{\eta_p}{2} Tr(R_{WX}) \frac{(H_p + \varsigma_v^2)(\sigma^2 + \varsigma_v^2 + H_p)^{3/2}}{(\sigma^2 + 2\varsigma_v^2 + 2H_p)^{3/2}} \quad (40)$$

Theorem 1. In a Gaussian noise environment and with the same step-size, the proposed nonlinear Hammerstein adaptive filter under MCC criterion has a smaller steady-state EMSE than under MSE criterion. As the kernel width increases, their values of the steady-state EMSE will become almost identical.

Proof. It can be shown that [34]

$$H_{p-MSE} = \frac{\eta_p Tr(R_{WX}) \varsigma_v^2}{2 - \eta_p Tr(R_{WX})} \quad (41)$$

where H$_{p-MSE}$ denotes the steady-state EMSE under MSE criterion. From Equation (40), we have

$$H_p = \frac{\varepsilon\eta_p Tr(R_{WX})\varsigma_v^2}{2-\varepsilon\eta_p Tr(R_{WX})} \qquad (42)$$

where $\varepsilon = \frac{(\sigma^2+\varsigma_v^2+H_p)^{3/2}}{(\sigma^2+2\varsigma_v^2+2H_p)^{3/2}}$. Since ε<1, it holds

$$H_p < H_{p-MSE} \qquad (43)$$

Further, as σ → ∞, we have H$_p$ → H$_{p-MSE}$.

Case B. Non-Gaussian Noise

Taking the Taylor series expansion of $f(e(n))$ around $v(n)$ yields

$$f(e(n)) = f(e_p(n)+v(n))$$
$$= f(v(n))+f'(v(n))e_p(n)+\frac{1}{2}f''(v(n))e_p^2(n)+h.o.t \qquad (44)$$

with

$$f'(v(n)) = \exp\left(\frac{-v^2(n)}{2\sigma^2}\right)\left(1-\frac{v^2(n)}{\sigma^2}\right)$$
$$f''(v(n)) = \exp\left(\frac{-v^2(n)}{2\sigma^2}\right)\left(\frac{v^3(n)}{\sigma^4}-\frac{3v(n)}{\sigma^2}\right) \qquad (45)$$

Under the assumptions (A) and (B), we get [34]

$$\lim_{n\to\infty} E\left[f(e(n))\cdot e_p(n)\right] \approx E\left[f'(v(n))\right]H_p \qquad (46)$$

$$\lim_{n\to\infty} E\left[f^2(e(n))\right] \approx E\left[f^2(v(n))\right]+E\left[f(v(n))f''(v(n))+\left|f'(v(n))\right|^2\right]H_p \qquad (47)$$

Substituting Equations (46) and (47) into Equation (36), we have

$$H_p = \frac{\eta_p Tr(R_{WX})E\left[f^2(v(n))\right]}{2E\left[f'(v(n))\right]-\eta_p Tr(R_{WX})E\left[f(v(n))f''(v(n))+\left|f'(v(n))\right|^2\right]} \qquad (48)$$

Further, substituting Equation (45) into Equation (48), we obtain

$$H_p = \frac{\eta_p Tr(R_{WX})E\left[\exp\left(\frac{-v^2(n)}{\sigma^2}\right)v^2(n)\right]}{2E\left[\exp\left(\frac{-v^2(n)}{2\sigma^2}\right)\left(1-\frac{v^2(n)}{\sigma^2}\right)\right]-\eta_p Tr(R_{WX})E\left[\exp\left(\frac{-v^2(n)}{\sigma^2}\right)\left(1+\frac{2v^4(n)}{\sigma^4}-\frac{5v^2(n)}{\sigma^2}\right)\right]} \quad (49)$$

When only the linear filter with parameter vector $\mathbf{w}(n)$ is adapted, we get

$$2\cdot\lim_{n\to\infty}E\left[f(e(n))\cdot e_w(n)\right]=\eta_p Tr(R_{XP})\lim_{n\to\infty}E\left[f^2(e(n))\right] \quad (50)$$

where $e_w(n)\approx\tilde{w}^T(n)X^T(n)p(n)$, $\tilde{w}(n)=w^*-w(n)$. For Gaussian noise case, we obtain

$$H_w=\frac{\eta_w}{2}Tr(R_{XP})\frac{(H_w+\varsigma_v^2)(\sigma^2+\varsigma_v^2+H_w)^{3/2}}{(\sigma^2+2\varsigma_v^2+2H_w)^{3/2}} \quad (51)$$

In non-Gaussian environments, we have

$$H_w=\frac{\eta_w Tr(R_{XP})E\left[\exp\left(\frac{-v^2(n)}{\sigma^2}\right)v^2(n)\right]}{2E\left[\exp\left(\frac{-v^2(n)}{2\sigma^2}\right)\left(1-\frac{v^2(n)}{\sigma^2}\right)\right]-\eta_w Tr(R_{XP})E\left[\exp\left(\frac{-v^2(n)}{\sigma^2}\right)\left(1+\frac{2v^4(n)}{\sigma^4}-\frac{5v^2(n)}{\sigma^2}\right)\right]} \quad (52)$$

Theorem 2. The H_{pw} satisfies the following condition

$$H_{pw}\geq H_p+H_w \quad (53)$$

Proof. Using Equation (31), we derive

$$\begin{aligned}e_{pw}(n)&=w^{*T}X^T(n)p^*-w^T(n)X^T(n)p(n)\\&=\left(w^{*T}X^T(n)p^*-w^{*T}(n)X^T(n)p(n)\right)+\left(w^{*T}X^T(n)p(n)-w^T(n)X^T(n)p(n)\right)\\&\approx w^T(n)X^T(n)\tilde{p}(n)+\tilde{w}^T(n)X^T(n)p(n)\\&=e_p(n)+e_w(n)\end{aligned}$$

$$(54)$$

It follows that

$$\begin{aligned}H_{pw}&=\lim_{n\to\infty}E\left[e^2_{pw}(n)\right]\approx\lim_{n\to\infty}E\left[\left(e_p(n)+e_w(n)\right)^2\right]\\&=\lim_{n\to\infty}E\left[e_p^2(n)\right]+2\lim_{n\to\infty}E\left[e_p(n)e_w(n)\right]+\lim_{n\to\infty}E\left[e_w^2(n)\right]\\&=H_p+H_w+2H_{cross}\end{aligned}$$

$$(55)$$

where $H_{cross}=\lim_{n\to\infty}E\left[e_p(n)e_w(n)\right]$ stands for the *cross-EMSE* and $H_{cross}\geq 0$ ($H_{cross}=0$ when $e_p(n)$ and $e_w(n)$ are statistically independent and zero mean) [7]. Therefore, $H_{pw}\geq H_p+H_w$, which completes the proof.

4. SIMULATION RESULTS

Now, we present simulation results to demonstrate the performance of the Hammerstein adaptive filtering under MCC. In order to show the performance of the proposed algorithm in non-Gaussian noises, we adopt the alpha-stable distribution to generate the disturbance noise, whose characteristic function is [32,46]

$$f(t) = \exp\{j\delta t - \gamma |t|^{\alpha} [1 + j\beta \operatorname{sgn}(t) S(t, \alpha)]\} \tag{56}$$

in which

$$S(t, \alpha) = \begin{cases} \tan \dfrac{\alpha \pi}{2} & \text{if } \alpha \neq 1 \\ \dfrac{2}{\pi} \log |t| & \text{if } \alpha = 1 \end{cases} \tag{57}$$

where $\alpha \in (0, 2]$ denotes the characteristic factor, $-\infty < \delta < +\infty$ is the location parameter, $\beta \in [-1, 1]$ stands for the symmetry parameter, and $\gamma > 0$ is the dispersion parameter. The characteristic factor α measures the tail heaviness of the distribution. The smaller α is, the heavier the tail is. In addition, γ measures the dispersion of the distribution. The distribution is symmetric about its location δ when $\beta = 0$. Such a distribution is called a *symmetric alpha-stable* (SαS) distribution. The parameters vector of the noise model is defined as $V = (\alpha, \beta, \gamma, \delta)$.

In the simulations below, the input signal considered is a colored signal obtained from the following equation:

$$x(n) = ax(n-1) + \sqrt{1 - a^2}\,\xi(n) \tag{58}$$

with a = 0.95, and $\xi(n)$ being a white Gaussian signal of unit variance. In addition, the coefficient vectors are initialized with the first coefficient equal to 1 and the others equal to zero [7].

4.1. Experiment 1

First, we consider an unknown Hammerstein system with parameter vectors $\mathbf{p}^* = [1, 0.6]$, $\mathbf{w}^* = [1, 0.6, 0.1, -0.2, -0.06, 0.04, 0.02, -0.03, -0.02, 0.01]$. Thus, M = 2, N = 10. The kernel width σ is 1.0. The noise vector V is set at (1.2, 0, 0.6, 0), and the noise signal is shown in Figure 3. Simulation results are averaged over 100 independent Monte Carlo runs, and in each simulation, 15,000 iterations are run to ensure the algorithm will reach the steady state, and the steady-state MSE is obtained as an average over the last 2000 iterations. The step-sizes are set at $\mu_p = \mu_w = 0.005$ and $\mu_p = 0.01$,

$\mu_w = 0.01$ for MSE and MCC, respectively. Figure 4 shows the average convergence curves under MCC and MSE. As we can see, the Hammerstein adaptive filtering under MCC criterion achieves faster convergence speed and lower steady-state testing MSE than under MSE criterion. Here the testing MSE is evaluated on a test set with 100 samples.

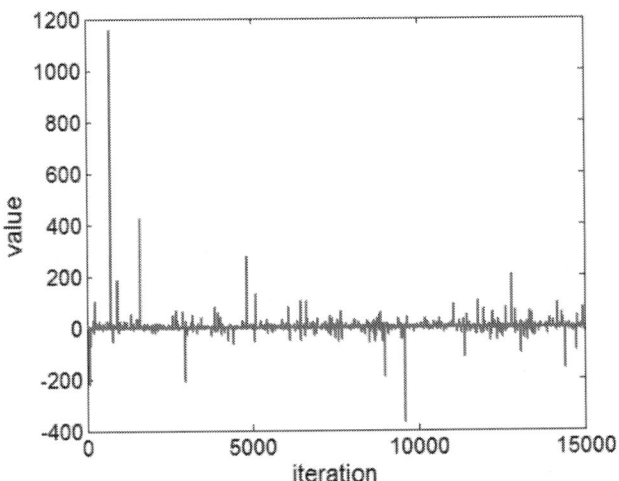

Figure 3. A typical sequence of the alpha-stable noise with V = (1.2, 0, 0.6, 0).

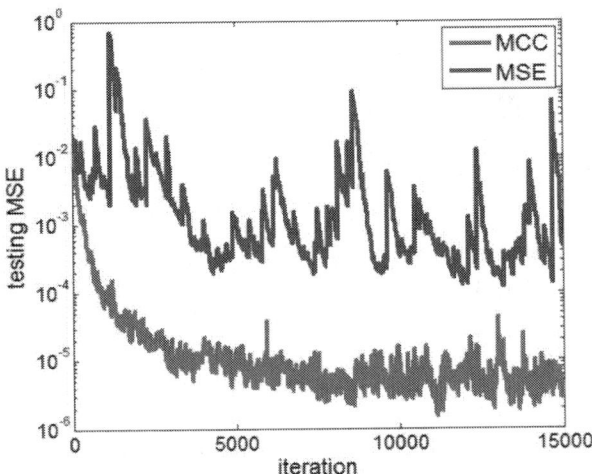

Figure 4. Convergence curves under maximum correntropy criterion (MCC) and mean square error (MSE) (for unknown system with polynomial nonlinearity).

Second, we investigate the performance of the algorithms with different noise parameters. The steady-state MSEs with different γ (0.2, 0.4, 0.6, 0.8, 1.0, 1.2, 1.4, 1.6) and different α (0.2, 0.4, 0.6, 0.8, 1, 1.2, 1.4, 1.6, 1.8, 2.0) are shown in Figure 5and Figure 6, respectively. We observe: (1) In most cases, the new algorithm performs better and achieves a lower steady-state MSE compared with the Hammerstein adaptive filtering under MSE criterion; (2) When α is close to 2.0, the Hammerstein adaptive filtering under MSE criterion can achieve better performance than under MCC criterion. The main reason for this is that, when α ≈ 2.0, the noise will be approximately Gaussian. Simulation results suggest that the proposed algorithm is particularly useful for identifying a Hammerstein system in non-Gaussian noises.

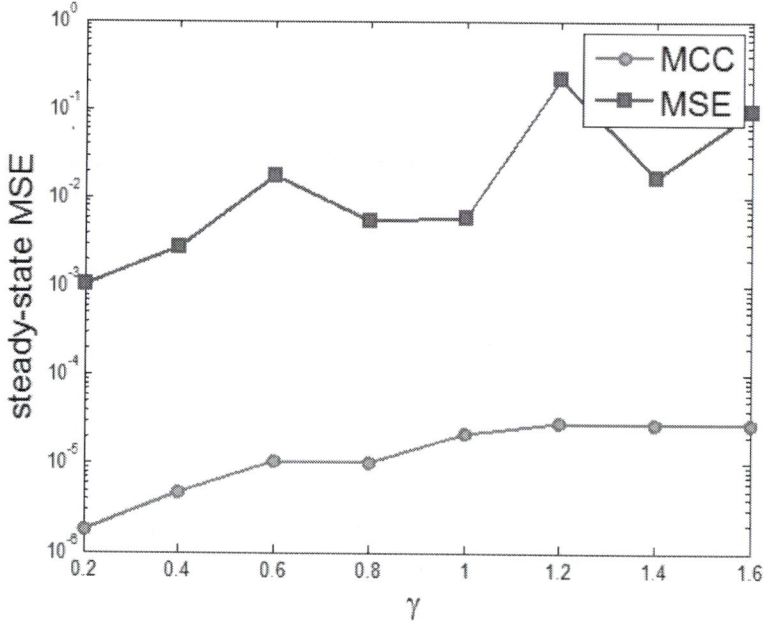

Figure 5. Steady-state mean square error (MSE) with different γ (α = 1.2).

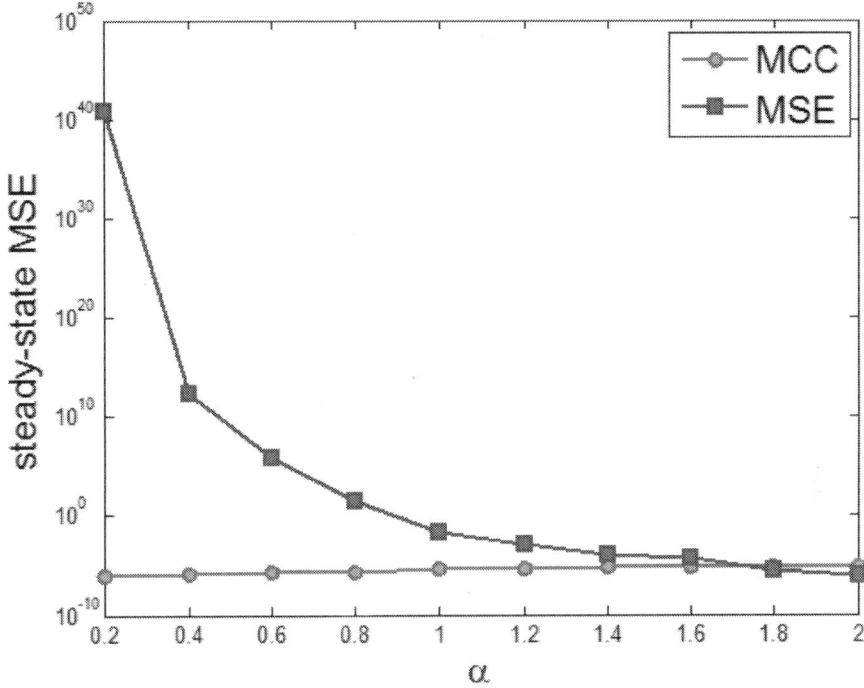

Figure 6. Steady-state mean square error (MSE) with different α (γ = 0.6).

4.2. Experiment 2

The second experiment is drawn from [47]. The nonlinear dynamic system is composed of two blocks. The first block is a non-polynomial nonlinearity

$$s(n) = \sqrt[3]{x(n)} \qquad (59)$$

while the second block is an FIR filter with weight vector

$$h = \begin{bmatrix} 1 & 0.75 & 0.5 & 0.25 & 0 & -0.25 \end{bmatrix}^T \qquad (60)$$

The noise vector V is set at (1.0, 0, 0.8, 0) (see Figure 7 for a typical sequence of the noise), and the polynomial order M and the FIR memory size N are set at 3 and 6, respectively. Simulation results are averaged over 50 independent Monte Carlo runs, and in each simulation, 30,000 iterations are run to ensure the algorithm will reach the steady state, and the steady-state MSE is obtained as an average over the last 2000 iterations. The testing MSE is evaluated on a test set with 100 samples. Figure 8 demonstrates the convergence curves under MCC and MSE. For both adaptive filtering algorithms, the step sizes are set as $\mu_1 = 0.005$, $\mu_2 = 0.015$. It can be seen that, the Hammerstein adaptive filter under MCC criterion performs better (say, with faster convergence speed and smaller mismatch error) than under MSE criterion.

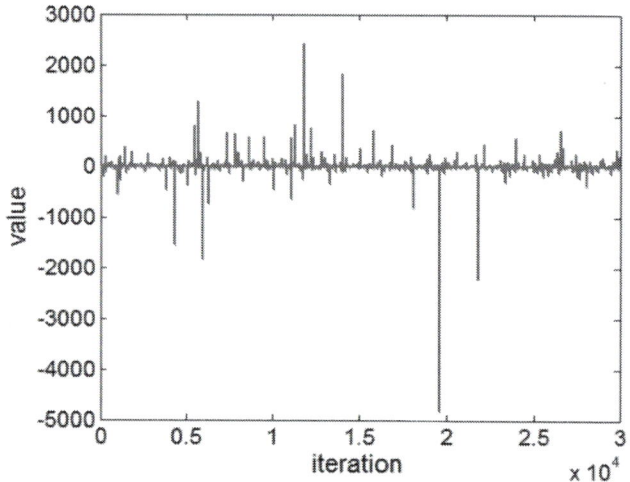

Figure 7. A typical sequence of the alpha-stable noise with V = (1.0, 0, 0.8, 0).

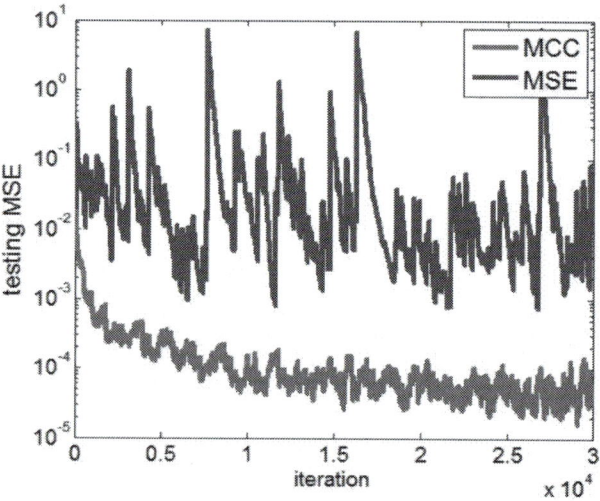

Figure 8. Convergence curves under maximum correntropy criterion (MCC) and mean square error (MSE) (for unknown system with non-polynomial nonlinearity).

Finally, we show the steady-state performance of the algorithms with different kernel widths σ (0.01, 1.0, 2.0, 3.0, 4.0, 5.0). Simulation results are shown in Figure 9. As we can see, the kernel width has significant influence on the performance of the proposed algorithm. In this example, the lowest steady-state MSE is obtained when σ = 1.0.

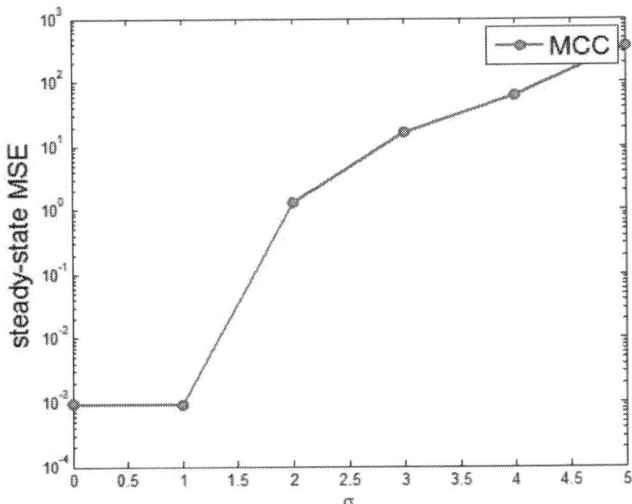

Figure 9. Steady-state mean square error (MSE) with different kernel widths.

5. CONCLUSIONS

The MCC has been successfully applied in domains of machine learning and signal processing due to its strong robustness in impulsive non-Gaussian situations. In this work, we develop a robust Hammerstein adaptive filter under MCC criterion. Different from the traditional Hammerstein adaptive filtering algorithms, the new algorithm use the MCC instead of the well-known MSE as the adaptation criterion, which can achieve desirable performance especially in impulsive noises. Based on [7,31], we carry out the convergence analysis, and obtain some important theoretical results. Simulation examples confirm the excellent performance of the proposed algorithm. How to verify the derived theoretical results is an interesting topic for future study.

ACKNOWLEDGMENTS

This work was supported by 973 Program (No. 2015CB351703) and National Natural Science Foundation of China (No. 61372152, No. 61271210).

AUTHOR CONTRIBUTIONS

Zongze Wu derived the algorithm and wrote the draft; Siyuan Peng proved the convergence properties and performed the simulations. Badong Chen proposed the main idea and polished the language; Haiquan Zhao was in charge of technical checking. All authors have read and approved the final manuscript.

REFERENCES

1. Ljung, L. *System Identification—Theory for the User*, 2nd ed.; Prentice Hall: Upper Saddle River, NJ, USA, 1999.

2. Sayed, A.H. *Fundamentals of Adaptive Filtering*; Wiley: Hoboken, NJ, USA, 2003.

3. Wiener, N. *Nonlinear Problems in Random Theory*; MIT Press: Cambridge, MA, USA, 1958.

4. Scarpiniti, M.; Comminiello, D.; Parisi, R.; Uncini, A. Nonlinear spline adaptive filtering. *Signal Process.* **2013**, *93*, 772–783.

5. Bai, E.W. Frequency domain identification of Wiener models. *Automatica* **2003**, *39*, 1521–1530.

6. Ogunfunmi, T. *Adaptive Nonlinear System Identification: The Volterra and Wiener Model Approaches*; Springer: Berlin/Heidelberg, Germany, 2007.

7. Scarpiniti, M.; Comminiello, D.; Parisi, R.; Uncini, A. Hammerstein uniform cubic spline adaptive filters: Learning and convergence properties. *Signal Process.* **2014**, *100*, 112–123.

8. Ortiz Batista, E.L.; Seara, R. A new perspective on the convergence and stability of NLMS Hammerstein filters. In Proceedings of the IEEE 8th International Symposium on Image and Signal Processing and Analysis (ISPA), Trieste, Italy, 4–6 September 2013; pp. 343–348.

9. Bai, E.W.; Li, D. Convergence of the iterative Hammerstein system identification algorithm. *IEEE Trans. Autom. Control* **2004**, *49*, 1929–1940.

10. Umoh, I.; Ogunfunmi, T. An Affine-Projection-Based Algorithm for Identification of Nonlinear Hammerstein Systems. *Signal Process.* **2010**, *90*, 2020–2030.

11. Umoh, I.; Ogunfunmi, T. An Adaptive Nonlinear Filter for System Identification. *EURASIP J. Adv. Signal Process.* **2009**, *2009*.

12. Umoh, I.; Ogunfunmi, T. An adaptive algorithm for Hammerstein filter system identification. In Proceedings of the EURASIP European Signal

Processing Conference, Lausanne, Switzerland, 25–29 August 2008; pp. 1–5.

13. Greblicki, W. Continuous time Hammerstein system identification. *IEEE Trans. Autom. Control* **2000**, *45*, 1232–1236.

14. Jeraj, J.; Mathews, V.J. Stochastic mean-square performance analysis of an adaptive Hammerstein filter. *IEEE Trans. Signal Process.* **2006**, *54*, 2168–2177.

15. Voros, J. Iterative algorithm for parameter identification of Hammerstein systems with two-segment nonlinearities. *IEEE Trans. Autom. Control* **1999**, *44*, 2145–2149.

16. Greblicki, W. Stochastic approximation in nonparametric identification of Hammerstein systems. *IEEE Trans. Autom. Control* **2002**, *47*, 1800–1810.

17. Ding, F.; Liu, X.P.; Liu, G. Identification methods for Hammerstein nonlinear systems. *Dig. Signal Process.* **2011**, *21*, 215–238.

18. Jeraj, J.; Matthews, V.J. A stable adaptive Hammerstein filter employing partial orthogonalization of the input signals. *IEEE Trans. Signal Process.* **2006**, *54*, 1412–1420.

19. Nordsjo, A.E.; Zetterberg, L.H. Identification of certain time-varying nonlinear Wiener and Hammerstein systems. *IEEE Trans. Signal Process.* **2001**, *49*, 577–592.

20. Atiya, A.; Parlos, A. Nonlinear system identification using spatiotemporal neural networks. In Proceedings of the International Joint Conference on Neural Networks, Baltimore, MD, USA, 7–11 June 1992; pp. 504–509.

21. Volterra, V. *Theory of Functionals and of Integral and Integro-Differential Equations*; Courier Corporation: North Chelmsford, MA, USA, 2005.

22. Principe, J.C.; Liu, W.; Haykin, S. *Kernel Adaptive Filtering: A Comprehensive Introduction*; Wiley: Hoboken, NJ, USA, 2011.

23. Chen, B.; Zhao, S.; Zhu, P.; Principe, J.C. Quantized kernel least mean square algorithm. *IEEE Trans. Neural Netw. Learn. Syst.* **2012**, *23*, 22–32.

24. Chen, B.; Zhao, S.; Zhu, P.; Principe, J.C. Quantized kernel recursive least squares algorithm. *IEEE Trans. Neural Netw. Learn. Syst.* **2013**, *24*, 1484–1491.

25. Chen, B.; Zhao, S.; Zhu, P.; Principe, J.C. Mean square convergence analysis of the kernel least mean square algorithm. *Signal Process.* **2012**, *92*, 2624–2632.

26. Bai, E.W. A blind approach to Hammerstein model identification. *IEEE Trans. Signal Process.* **2002**, *50*, 1610–1619.

27. Stenger, A.; Kellermann, W. Adaptation of a memoryless preprocessor for nonlinear acoustic echo cancelling. *Signal Process.* **2000**, *80*, 1747–1760.

28. Shi, K.; Ma, X.; Zhou, G.T. An efficient acoustic echo cancellation design for systems with long room impulses and nonlinear loudspeakers. *Signal Process.* **2009**, *89*, 121–132.

29. Scarpiniti, M.; Comminiello, D.; Parisi, R.; Uncini, A. Comparison of Hammerstein and Wiener systems for nonlinear acoustic echo cancelers in reverberant environments. In Proceedings of the 17th International Conference on Digital Signal Processing (DSP), Corfu, Greece, 6–8 July 2011; pp. 1–6.

30. Kailath, T.; Sayed, A.H.; Hassibi, B. *Linear Estimation*; Prentice Hall: Upper Saddle River, NJ, USA, 2000.

31. Plataniotis, K.N.; Androutsos, D.; Venetsanopoulos, A.N. Nonlinear filtering of non-Gaussian noise. *J. Intell. Robot. Syst.* **1997**, *19*, 207–231.

32. Weng, B.; Barner, K.E. Nonlinear system identification in impulsive environments. *IEEE Trans. Signal Process.* **2005**, *53*, 2588–2594.

33. Chen, B.; Zhu, Y.; Hu, J.; Principe, J.C. *System Parameter Identification: Information Criteria and Algorithms*; Elsevier: Amsterdam, The Netherlands, 2013.

34. Chen, B.; Xing, L.; Liang, J.; Zheng, N.; Principe, J.C. Steady-state Mean-square Error Analysis for Adaptive Filtering under the Maximum Correntropy Criterion. *IEEE Signal Process. Lett.* **2014**, *21*, 880–884.

35. Principe, J.C. *Information Theoretic Learning: Renyi's Entropy and Kernel Perspectives*; Springer: New York, NY, USA, 2010.

36. Singh, A.; Principe, J.C. Using Correntropy as a cost function in linear adaptive filters. In Proceedings of the IEEE International Joint Conference on Neural Networks (IJCNN), Atlanta, GA, USA, 14–19 June 2009; pp. 2950–2955.

37. Chen, B.; Principe, J.C. Maximum correntropy estimation is a smoothed MAP estimation. *IEEE Signal Process. Lett.* **2012**, *19*, 491–494.

38. Zhao, S.; Chen, B.; Principe, J.C. Kernel adaptive filtering with maximum correntropy criterion. In Proceedings of the IEEE International Joint Conference on Neural Networks (IJCNN), Killarney, Ireland, 31 July–5 August 2011; pp. 2012–2017.

39. Ogunfunmi, T.; Paul, T. The Quaternion Maximum Correntropy Algorithm. *IEEE Trans. Circuits Syst. -II (TCAS-II)* **2015**, *62*, 598–602.

40. Mandic, D.P.; Hanna, A.I.; Razaz, M. A normalized gradient descent algorithm for nonlinear adaptive filters using a gradient adaptive step size. *IEEE Signal Process. Lett.* **2001**, *8*, 295–297.

41. Al-Naffouri, T.Y.; Sayed, A.H. Adaptive filters with error non-linearities: Mean-square analysis and optimum design.*EURASIP J. Appl. Signal Process.* **2001**, *4*, 192–205.

42. Lin, B.; He, R.; Wang, X.; Wang, B. The steady-state mean-square error analysis for least mean p-order algorithm.*IEEE Signal Process. Lett.* **2009**, *16*, 176–179.

43. Yousef, N.R.; Sayed, A.H. A unified approach to the steady-state and tracking analysis of adaptive filters. *IEEE Trans. Signal Process.* **2001**, *49*, 314–324.

44. Duttweiler, D.L. Adaptive filter performance with nonlinearities in the correlation multiplier. *IEEE Trans. Acoust. Speech Signal Process.* **1982**, *30*, 578–586.

45. Mathews, V.J.; Cho, S.H. Improved convergence analysis of stochastic gradient adaptive filters using the sign algorithm. *IEEE Trans. Acoust. Speech Signal Process.* **1987**, *35*, 450–454.

46. Shao, M.; Nikias, C.L. Signal processing with fractional lower order moments: Stable processes and their applications.*Proc. IEEE* **1993**, *81*, 986–1010.

47. Hasiewicz, Z.; Pawlak, M.; Sliwinski, P. Nonparametric identification of nonlinearities in block-oriented systems by orthogonal wavelets with compact support. *IEEE Trans. Circuits Syst. I Regul.* **2005**, *52*, 427–442.

Chapter 5

Application of Adaptive Noise Cancellation in Transabdominal Fetal Heart Rate Detection Using Photoplethysmography

Kok Beng Gan[1], Edmond Zahedi[2], Mohd. Alauddin[1] and Mohd. Ali[1]

[1] Institute of Space Science, National University of Malaysia, Malaysia
[2] School of Electrical Engineering, Sharif University of Technology, Iran

1. INTRODUCTION

Monitoring the fetal heart rate (FHR) throughout pregnancy empowers the clinician to diagnose fetal well being, characterize fetal development and detect abnormality (Freeman et al. 2003). A non-invasive and low cost system would enable monitoring of normal pregnancies and promote large population studies of fetal physiological development (Freeman et al. 2003). FHR monitoring is an ongoing observation of human fetal physiology. The expected outcome of this early detection is a reduced risk of fetal morbidity and mortality (Philip et al. 2002).

Currently, Doppler ultrasound has been extensively used for FHR detection and obstetric purposes (Hershkovitz et al. 2002), where the standard pre-delivery test of fetal health is the fetal non-stress test (NST). These tests are routinely performed at the hospital where continuous-wave instruments are more popular than pulsed ones. The use of Doppler ultrasound in the first trimester is generally not recommended as a routine (Hershkovitz et al. 2002) as it may increases the occurrence of intrauterine growth restriction. Besides that, the FHR measurements using Doppler ultrasound are not always reliable (Karlsson et al. 2000) due to the complexity of the Doppler signal and the effects of fetal and maternal breathing. An alternative to ultrasound is using the fetal electrocardiogram (FECG). In direct (invasive) fetal electrocardiogram (FECG), the FHR could be obtained by attaching scalp electrode to the fetal scalp after the rupture of the membrane (Khandpur 2004). During invasive FECG recording the uterus may be perforated leading to its infection, besides possible scalp injuries to the fetus (Khandpur 2004). The other approach is non-invasive FECG but FECG signals have a low (signal to noise ratio) SNR due to the interference from noise, maternal electrocardiogram (MECG) and electromyogram (EMG). The application of non-invasive FECG requires multiple leads and advanced digital signal processing techniques (Najafabadi et al. 2006). It is worth mentioning that commercial devices operating on non-invasive FECG are not available at this moment.

Optical techniques has received a considerable attention in biomedical diagnostic and monitoring of biological tissues such as brain imaging, breast imaging and for fetal heart rate detection and oxygen saturation measurement due to its theoretical advantages in comparison with other modalities. Continuous wave near infrared (NIR) spectroscopy has been applied to trans-abdominal fetal pulse oximetry (Ramanujam et al. 2000; Chance 2005; Zourabian et al. 2000; Nioka et al. 2005; Vintzileos et al. 2005). The system consists of NIR sources (halogen lamps) and a photomultiplier as detection unit (Ramanujam et al. 2000; Chance 2005). The generated heat was justified by using cooling fans for the halogen lamps. Recently, trans-abdominal oxygen saturation (S_pO_2) in animal (Nioka et al. 2005) and human fetuses were successfully obtained in the laboratory (Vintzileos et al. 2005). However, the proposed techniques require high power (a total of 80 W optical power) and a relatively expensive detection unit (photo-multiplier).

In this project, we propose to design and develop a low-power optical FHR monitor. The signal of interest is the photoplethysmogram (PPG), which is generated when a beam of light is modulated by blood pulsations. PPG is a noninvasive technique for detecting blood volume changes in living tissue by optical means consisting of a light emitting diode (LED) and a photo-detector. One of the potential applications of the PPG technology is non-invasive fetal heart rate detection through the maternal abdomen. In this application, the light intensity is modulated by the mother as well as fetal blood circulation, producing a combined signal which needs to be separated via digital signal processing (DSP) techniques. The design of a fixed filter would not be adequate as the frequency spectrum of the noise (maternal PPG) overlaps with the desired signal (fetal PPG). The adaptive filter will automatically adjust its coefficients therefore achieve the high degree of noise rejection.

Such an approach - based on adaptive noise cancellation (ANC) - has been evaluated for extraction of the fetal heart-rate using PPG signals from the maternal abdomen. A simple optical model has been proposed in which the maternal and fetal blood pulsations result in emulated signals where the lower SNR limit (fetal to maternal) is -25 dB (Zahedi & Beng, 2008). It is shown that the RLS algorithm is capable to extract the peaks of the fetal PPG from these signals, corresponding to typical values of maternal and fetal tissues.

Subsequently, an optical fetal heart rate detection (OFHR) system has been designed and developed using low-cost, low-power IR light (890 nm with optical power < 68 mW) and a commercially available silicon photo-detector (Gan et al. 2009). Previous literature (Ramanujam et al. 2000; Chance 2005; Zourabian et al. 2000; Nioka et al. 2005; Vintzileos et al. 2005; Choe et al. 2003) shows that the Source-Detector separations depends on the type of sources and the photo-detectors implemented in their studies. Since in our work the developed instrument utilizes low optical power, the source-to-detector separation plays an important role as it affects the detectivity of the photo-detector. This chapter discusses the selection of S-D separation for the OFHR system based on the ANC limit and photo-detector's noise. The implementation of the ANC algorithm in OFHR system is also discussed and the clinical trial results are also reported.

2. MATERIALS AND METHODS

2.1. Adaptive Noise Cancellation

Conventional digital signal processing techniques do exist to extract a desired biomedical signal from a mixed signal which is usually contaminated by unwanted noises. Adaptive filters are used for non-stationary signals where a sample-by-sample adaptation process is required (Vaseghi, 2000; Widrow et al., 1975). Applications of adaptive filtering include multi-channel noise reduction, radar or sonar signal processing, channel equalization for cellular mobile phones, echo cancellation and low delay speech coding. This section discussed the concept of the adaptive filtering, adaptive algorithm and the Recursive Least Square (RLS) algorithm.

2.1.1. Concept Of Adaptive Filtering

Adaptive filters consist of two distinct parts: a digital filter and the corresponding adaptive algorithm, used to adjust the coefficients of the filter (Figure 1). In these algorithms, the error signal $e(n)$ defined as the difference between the output of the filter ($y(n)$) and a primary input signal ($d(n)$), is minimized according to a least squares error criterion (Ifeachor & Jervis, 2002).

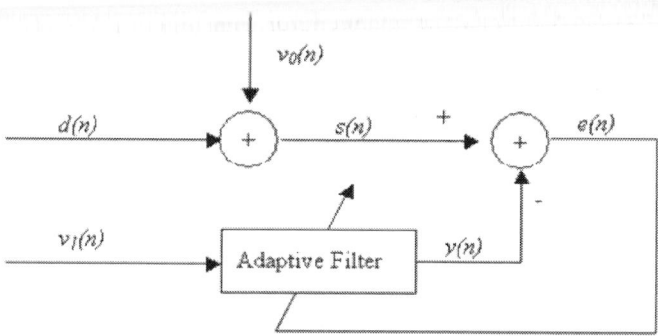

Figure 1. ANC system

The desired signal $d(n)$ (Figure 1) is contaminated by an uncorrelated noise signal $v_0(n)$, where n is the running time index. The result $d(n) + v_0(n)$ is the primary measurement signal $s(n)$. The reference input, $v1(n)$ is only correlated with $v_0(n)$ and fed to an adaptive FIR filter. The output of the FIR adaptive filter $y(n)$ is subtracted from the primary input $s(n)$ to produce the error signal $e(n)$:

$$e(n) = d(n) + v_0(n) - y(n) \tag{1}$$

The adaptive filter uses $e(n)$ to adjust its own impulse response to produce an output $y(n)$ as close a replica as possible to $v_0(n)$. Squaring and applying the expectation operator to both sides of Equation 1:

$$E\{e^2(n)\} = E\{d^2(n)\} + E\{(v_0(n) - y(n))^2\} - 2E\{d(n)(v_0(n) - y(n))\} \tag{2}$$

$d(n)$ being uncorrelated with $v_0(n)$ and $v_1(n)$, $E\{d(n)(v0(n)-y(n))\}=$ 0. Therefore Equation 2 can be simplified:

$$E\{e^2(n)\} = E\{d^2(n)\} + E\{(v_0(n) - y(n))^2\} \tag{3}$$

An iterative procedure minimizes $E\{e^2(n)\}$, which will occur when $y(n) = v_0(n)$ (ideal situation) producing $e(n) = d(n)$.

2.1.2. Adaptive Algorithm

In most adaptive systems, the digital filter in Figure 2 is realized using a transversal or finite impulse response (FIR) structure. The FIR structure is the most widely used because of its simplicity and stability.

A mth-order adaptive transversal filter is a linear time varying discrete-time system that can be represented by:

$$y(n) = \sum_{i=0}^{m-1} w_i(n)v_1(n-i)$$ (4)

where $wi(n)$ is the adjustable weight and $v1(n)$ and $y(n)$ are the input and output of the filter. The filter output is a time varying linear combination of the past input (Figure 2).

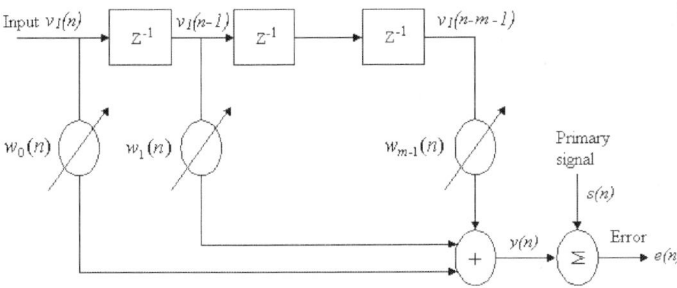

Figure 2. Illustration of the configuration of an adaptive filter

Adaptive algorithm are used to adjust the coefficient of the digital filter (Figure 2) such that the error signal $e(n)$, is minimized according to the mean square error and least squares error criterion (Ifeachor & Jervis, 2002). Common adaptation algorithms are least mean square (LMS) and the RLS. The RLS algorithm minimizes the sum of the square of the error whereas the LMS algorithm minimizes the mean square error. In terms of the computational and storage requirements, the LMS algorithm is the most efficient and does not suffer from the numerical instability problem (Ifeachor & Jervis, 2002). However, the recursive least square (RLS) algorithm has superior convergence properties (Ifeachor & Jervis, 2002). It is suitable for offline processing where computational requirement is not an issue.

2.1.3. Linear Least-Square Error Estimation

The principle of least-squares (LS) was introduced by the German mathematician Carl Friedrich Gauss, who used it to determine the orbit of the asteroid ceres in 1821 by formulating the estimation problem as an optimization problem (Manolakis et al., 2005).

The least-square approach provides a mechanism for designing fixed filters when the properties of the signal source are known. More importantly, it provides a vehicle for adaptive filter design that can operate in an environment of changing signal properties. The source signal is modeled as the output of a linear discrete-time system with parameters which are either known for the fixed algorithm or unknown in the adaptive case. Noise added to the observations completes the signal description. The least-square algorithm is then required to do the "best" filtering of the signal, employing as much of the priori signal and noise models as is known. If these priori properties are unknown, then the LS algorithm is required to identify the changed conditions and to adapt its parameters to the new signal environment.

The basic idea of the LS method is shown in Figure 3. An output signal, s(n) measured at the discrete time, n in response to a set of input signal, v1(n). The input and output signals are related by the simple regression model.

$$s(n) = \sum_{i=0}^{m-1} w_i(n)v_1(n) + e(n) \tag{5}$$

where e(n) is the measurement errors and wi(n) is the adjustable weight with mth order.

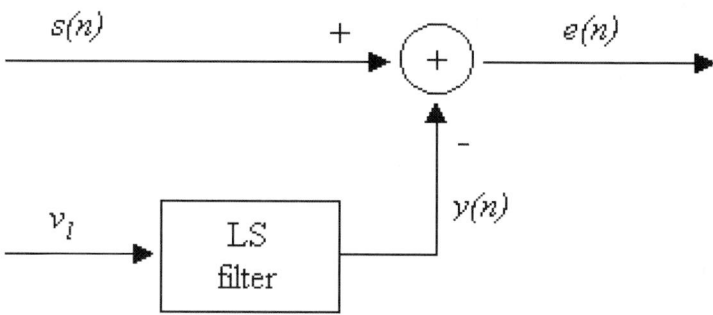

Figure 3. An illustration of the basic idea of the LS method

The estimation error is defined as

$$e(n) = s(n) - \sum_{i=0}^{m-1} w_i(n)v_1(n)$$

$$= s(n) - w^T v_1$$

(6)

where $v1 = [v1(n), v1(n-1),..., v1(n-m-1)]^T$ and $w = [w0(n), w1(n),..., wm-1(n)]^T$. The filter weight,$wi(n)$ are determined by minimizing the sum of the squared errors

$$E \triangleq \sum_{n=0}^{n-1} |e(n)|^2$$

(7)

that is, the energy of the signal.

To explore the relation between the filter coefficient, w, and the error signal, $e(n)$, Equation 6 can be written in matrix form for N samples measurement of the signals $[s(0), s(1),..., s(N–1)]$ and signals $[v1(n), v1(1),..., v1(N-1)]$ as

$$\begin{pmatrix} e(0) \\ e(1) \\ e(2) \\ \vdots \\ e(N-1) \end{pmatrix} = \begin{pmatrix} s(0) \\ s(1) \\ s(2) \\ \vdots \\ s(N-1) \end{pmatrix} - \begin{pmatrix} v_{10}(0) & v_{11}(0) & v_{12}(0) & \cdots & v_{1m-1}(0) \\ v_{10}(1) & v_{11}(1) & v_{12}(1) & \cdots & v_{1m-1}(1) \\ v_{10}(2) & v_{11}(2) & v_{12}(2) & \cdots & v_{1m-1}(2) \\ \vdots & \vdots & \vdots & \ddots & \vdots \\ v_{10}(N-1) & v_{11}(N-1) & v_{12}(N-1) & \cdots & v_{1m-1}(N-1) \end{pmatrix} \begin{pmatrix} w_0 \\ w_1 \\ w_2 \\ \vdots \\ w_{m-1} \end{pmatrix}$$

(8)

or more compactly as

$$E = s - Vw$$

(9)

where

$$\begin{aligned} e &\triangleq [e(0), e(1),...,e(N-1)]^T & \text{error data vector} (N \times 1) \\ s &\triangleq [s(0), s(1),...,s(N-1)]^T & \text{primary data vector} (N \times 1) \\ V &\triangleq [v_1(0), v_1(1),...,v_1(N-1)]^T & \text{input data matrix} (N \times m) \\ w &\triangleq [w_0, w_1,...,w_{m-1}]^T & \text{weight vector} (m \times 1) \end{aligned}$$

(10)

where $w1(n) \triangleq [w10(0), w11(1),..., w1m\ 1(n)]$. The energy of the error vector, that is the sum of squared elements of the squared error vector, is given by the inner vector product as:

$$e^T e = (s - Vw)^T (s - Vw)$$
$$= s^T s - s^T Vw - V^T w^T s + V^T w^T Vw$$

(11)

The gradient of the squared error function with respect to the filter coefficients is obtained by differentiating Equation 11 with respect to w as:

$$\frac{\partial e^T e}{\partial w} = -2s^T V + 2w^T V^T V$$

(12)

The filter coefficients are obtained by setting the gradient of the squared error function of Equation 12to zero and yield:

$$(V^T V)w = V^T s$$

(13)

or

$$w = (V^T V)^{-1} V^T s$$

(14)

Note that the matrix $V^T V$ is a time-averaged estimate of the autocorrelation matrix of the input signal, Ryy and the vector $V^T s$ is a time-averaged estimate of the cross-correlation vector of the input and the primary signals, ryx

2.1.4. Recursive Least Square Algorithm

The RLS algorithm is based on the least-square method (Ifeachor & Jervis, 2002; Haykin, 2002). In recursive implementations of the method of least squares, the computation is started with known initial conditions and use the information contained in new data samples to update the old estimates. The RLS adaptive filters are designed so that the updating of the coefficients is always achieved the minimization of the sum of the squared errors. The RLS adaptive algorithm for updating the coefficients of the FIR filter is superior to the LMS algorithm in convergence properties, eigen value sensitivity, and excess MSE. The price paid for this improvement is additional computational complexity.

The computation of **w** in Equation 14 requires time-consuming computation of the inverse matrix. With the RLS algorithm the estimate of w can be updated for each new set of data acquired without repeatedly solving the time-consuming matrix inversion directly. A suitable RLS algorithm can be obtained by exponential weighting the data to remove gradually the effect of old data on w and to allow the tracking of slowly varying signal characteristic.

The derivation of the RLS algorithm can be found in the report (Gan, 2009) and the RLS algorithm can be summarized as follows:

Input signals: $v1(n)$ and $d(n)$

Initial values:

$$\Phi_{yy}(n) = \delta^{-1}I$$

$$w(0) = w_I$$

For $n = 1, 2,...$, compute

1. Filter gain vector update :

$$k(n) = \frac{\lambda_f^{-1}\Phi_{yy}(n-1)v_1(n)}{1 + \lambda_f^{-1}v^T(n)\Phi_{yy}(n-1)v_1(n)} \tag{15}$$

2. Error signal equation:

$$e(n) = d(n) - w^T(n-1)v_1(n) \tag{16}$$

3. Filter coefficients adaptation:

$$w(n) = w(n-1) - k(n)e(n) \tag{17}$$

4. Inverse correlation matrix update:

$$\Phi_{yy}\left(n\right)=\lambda_{f}^{-1}\Phi_{yy}\left(n-1\right)-\lambda_{f}^{-1}k\left(n\right)v_{1}^{T}\left(n\right)\Phi_{yy}\left(n-1\right) \tag{18}$$

2.2. Photoplethysmography

Photoplethysmograph is an optoelectronic method for measuring and recording changes in the volume of body parts such as finger and ear lobes caused by the changes in volume of the arterial oxygenated blood, associated with cardiac contraction (Bronzino 2000). A sample of few normal periodic PPG pulse waves is shown in Figure 4, where the steep rise and dicrotic notch on the falling slope are clearly visible. When light travels through a biological tissue (earlobe or finger), it is absorbed by different absorbing substances. Primary absorbers are the skin pigmentation, bones and the arterial and venous blood. The characteristics of the PPG pulse are influenced by arterial ageing and arterial disease (Allen & Murray 2000).

The emitted light either red or infrared light emitting diode is detected by a photo-detector. The time varying signals of the detected signal is called PPG. The PPG signal contains AC and DC components: the AC component is mainly due to the arterial blood pulsation and the DC component comes from the non-pulsating arterial blood, venous blood and other tissues.

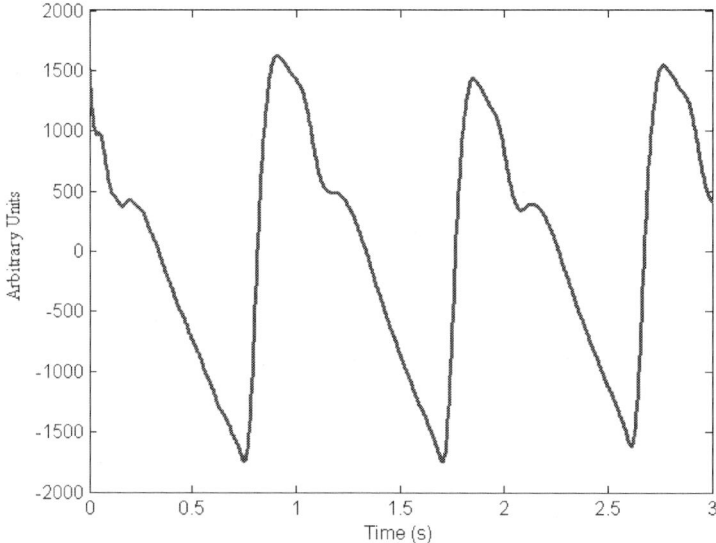

Figure 4. Typical PPG pulse wave signal acquired in our laboratory

The probes can be of two types, transmission or reflection. A transmission probe measures the amount of light that passes through the tissue as in a finger clip probe. The photodiode is located on the opposite side of the LED and the tissue is located between them. A reflectance probe measures the amount of light reflected to the probe. However, the detected light intensity of a reflectance probe is weaker than the transmission probe with the same source to detector separation.

In this application, transmission probes are not suitable due to the very long optical path that the light would have to travel to the photo-detector which is located opposite sides of the maternal abdomen (Zahedi & Beng, 2008). The reflectance probe becomes the method of choice where the photo-detector is placed on the same body surface (abdomen) making the measurement of abdominal PPG signal possible (Zahedi & Beng, 2008).

2.3. Photo-Detector Noise

When designing an optical instrument, the photo-detector is an essential component. Selection of an appropriate photo-detector resulted in better signal quality of the acquired signals. The noise floor of the photo-detector will determine the maximum S-D separation which is useful in the optical instruments.

Currently, the low noise photo-detector (from Edmund Optics Inc.) with noise equivalent power as low as 1.810^{-14} W/Hz$^{1/2}$ (0.051 cm^2) (W57-522, Edmund Optics, Inc.) and 8.610^{-14} W/Hz$^{1/2}$ (1.00 cm^2) (W57-513, Edmund Optics, Inc.). Noise equivalent power is the incident optical power required to produce a signal on the photo-detector that is equal to the noise when the SNR is equal to one. These silicon photo-detectors are then utilized in the following analysis.

The photo-detector can either operate in photovoltaic or photo-conductance condition. Photovoltaic operation offered a low noise system compared to the photo-conductance operation. Shot noise (due to the dark current) is the dominant noise component during photo-conductance operation. Small photo-detector's active area resulted in lower noise level compared to the large photo-detector's active area. Since strong scattering process for the human tissue dispersed the light in random fashion (Bronzino, 2000), large photo-detector's active area increases the probability of detecting photons that exit from the maternal layer. Therefore, photo-detector with 1 cm^2 area is proposed for the optical fetal heart rate instrument. This value has thus been used in the rest of this work. Table 1 showed the proposed silicon photo-detector's noise, *PNoise* during photovoltaic operation at various bandwidths. It shows that photo-detector's noise increases with its bandwidth.

TABLE 1. *PNoise* during photovoltaic operation at various bandwidths

Photo-detector area (cm2)	R (A/W)	Rsh min (M)	Bandwidth (Hz)	I_{PN} (A)
0.051	0.62	600	100	8.29 10-14
			1000	2.63 10-13
			10000	8.29 10-13
			100000	2.63 10-12
1	0.62	30	100	3.71 10-13
			1000	1.17 10-12
			10000	3.71 10-12
			100000	11-Oct

3. RESULTS AND DISCUSSIONS

This section discusses the determination of S-D separation based on the limit of ANC operation. Results obtained in previous work (Zahedi & Beng, 2008) encouraged us to take one step forward via practical implementation of the circuitry whereas digital synchronous detection is utilized to further enhance the SNR. The design and development of the OFHR system is described and results of the clinical trial are also reported.

3.1. Adaptive Noise Cancellation and The Limit of The Photo-Detector

Since the adaptive noise canceling limit is -34.7 dB, the photo-detector used in the optical fetal heart rate instrument must be able to detect fetal signal at this limit. By using Equation 19, the expected fetal optical power, *PF* at -34.7 dB is estimated and tabulated in Table 2.

$$10\log_{10}\left[\frac{P_F}{P_{M+am}}\right] = -34.7dB \qquad (19)$$

where P_F is the estimated fetal optical power, P_{M+am} is the optical power at photo-detector using Monte Carlo simulation and -34.7 dB is the limit of the ANC operation. These values were obtained through Monte-Carlo simulation using a three-layered tissue model (maternal, amniotic, and fetal) (Zahedi & Beng, 2008). Optical properties (scattering and absorption coefficients) of the tissue model as well as respective thicknesses were obtained from previous studies (Ramanujam et al. 2000; Gan, 2009), and simulation results were based on the launching of two million photons with 1 mW optical power. The detailed discussion of the Monte-Carlo simulation can be found in the previous report (Gan, 2009) and will not be further discussed here.

From Figure 5, when S-D separation larger than 4 cm (6 cm, 8 cm and 10 cm), the expected optical power is below the photo-detector noise level. At 2 cm and 4 cm source to detector separation, the expected fetal optical powers, 2293.99×10^{-12} W/cm^2 and 5.94×10^{-12} W/cm^2 respectively, are higher than the photo-detector's noise (1.17×10^{-12} W/cm^2) level. The photo-detector is assumed to be operated at the photovoltaic condition with 1000 Hz bandwidth and 1 cm^2 active area. Therefore, source to detector separation of 4 cm, which results in 70% of optical power from fetal layer, is suitable to use with this low noise photo-detector. At 890 nm and 4 cm source-detector separation, the receiver sensitivity is optimized by considering the limitation of the adaptive filter in FHR detection.

Table 2. Expected PF signal level (-34.7 dB) at various source to detector separation

Source to detector separation	Expected signal level, PM+am	Expected PF signal level of -34.7 dB
(cm)	(10-9)	(10-12)
2	6767.09	2293.99
4	17.53	5.94
6	0.31	0.11
8	0.37	0.13
10	0.09	0.03

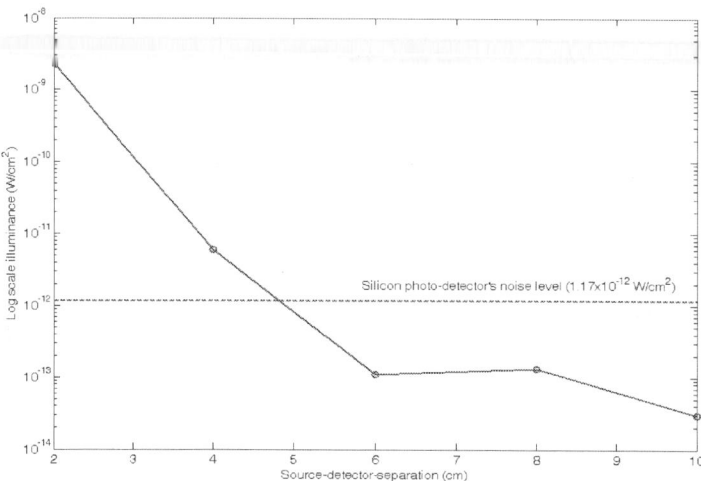

Figure 5. Estimated P_F (-34.7 dB) at 2.5 cm fetal depth

3.2. Implementation of ANC In Transabdominal Fetal Heart Rate Detection Using Ppg

In our work (Gan et al. 2009), a low-power optical technique is proposed based on the PPG to non-invasively estimate the FHR. A beam of LED light (<68 mW) is shone to the maternal abdomen and therefore modulated by the blood circulation of both mother and fetus whereas maximum penetration is achieved at a wavelength of 890 nm. This mixed signal is then processed by an adaptive filter with the maternal index finger PPG as reference input.

Figure 6 shows the optical fetal heart rate detection (OFHR) system block diagram whereas the implementation by using National Instrument hardware and LabVIEW software are illustrated inFigure 7 and Figure 8. In the OFHR system, the fetal probe (primary signal) is attached to the maternal abdomen using a Velcro belt to hold the IR-LED and photo-detector, separated by 4 cm. The reference probe is attached to the mother's index finger as generally practiced in pulse oximetry. As the selected IR-LED could only emit a maximum optical power of 68 mW, the OFHR system operates with an optical power less than the limit of 87 mW specified by the International Commission on Non-Ionizing Radiation Protection (ICNIRP) (International Commission on Non-Ionizing Radiation Protection, 2000). In order to modulate the IR-LED, the modulation signal is generated at a frequency of 725 Hz using software subroutine through a counter port (NI-USB 9474) to the LED driver (Figure 6).

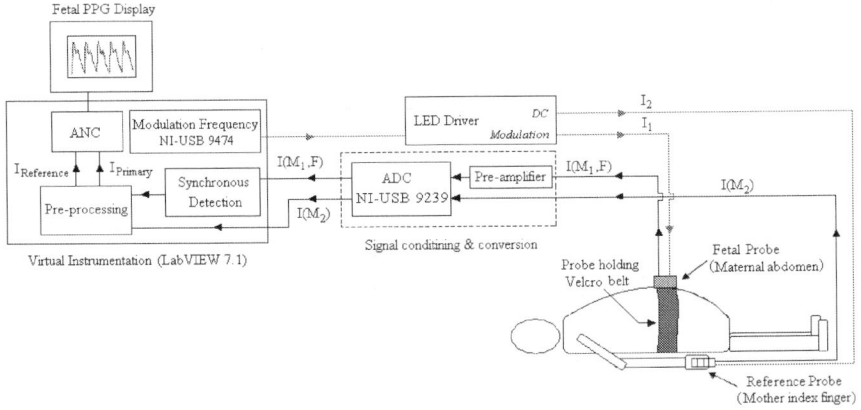

Figure 6. OFHR system block diagram showing the hardware modules have been implemented in LabVIEW.

The diffused reflected light from the maternal abdomen, detected by the low-noise photo-detector, is denoted as $I(M_1,\ F)$, where M_1 and F denote the contribution to the signal from the mother abdomen and fetus, respectively. A low-noise (6 nV/Hz$^{1/2}$) transimpedance amplifier is utilized to convert the detected current to a voltage level. The reference probe (mother's index finger) consists of an IR-LED and a solid-state photodiode with an integrated preamplifier. The signal from this probe is denoted as $I(M_2)$, where M_2 refers to the maternal contribution. Synchronous detection is not required at this channel as the finger photoplethysmogram has a high signal to noise ratio (SNR).

Detected signals from both probes are simultaneously digitized with a 24-bit resolution data acquisition card (NI-USB 9239, National Instruments, Inc.) at a rate of 5.5 kHz. The demodulation, digital filtering, and signal estimation are all performed in the digital domain. Software implementation consists of generating a modulation signal, a synchronous detection algorithm, down-sampling, high-pass filtering and ANC algorithm (Zahedi & Beng, 2008). The entire algorithm and part of the instrument have been implemented using Laboratory Virtual Instrumentation Engineering Workbench (LabVIEW 7.1, from National Instruments, Inc.). After pre-processing and applying the ANC algorithm (Figure 9), the fetal signal and the fetal heart rate are displayed. The FHR is found by estimating the prominent peak of the power spectral density using the Yule-Walker autoregressive (AR) method (order of 20).

Figure 7 shows the laboratory prototype and the graphical user interface of the OFHR system (inFigure 7, left) where the maternal index finger PPG (top), the abdominal PPG (middle) and the estimated fetal PPG (bottom) are presented. There are three types of selectable displays (Figure 8) namely digital synchronous or lock-in amplifier (LIA), ANC and heart rate trace. The purpose of the first two displays is to assist development and the third one (Figure 8) indicates FHR values versus time (clinical application). The user can either save the data to the personal computer for further analysis or just display it online.

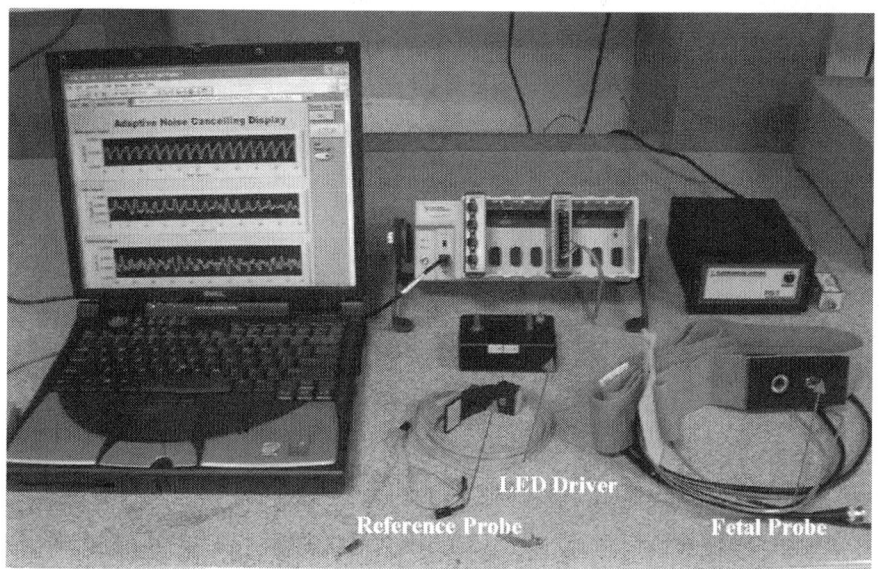

Figure 7. OFHR prototype

Finally, a total of 24 data sets were acquired from six subjects at 37±2 gestational weeks from the Universiti Kebangsaan Malaysia Medical Centre. This study was reviewed and approved by the University Ethical Committee and written consent was obtained from all patients who participated in this study after the procedure was clearly explained to them. The process for subject recruitment and data acquisition are complied with the rules and regulation as stated in Good Clinical Practice.

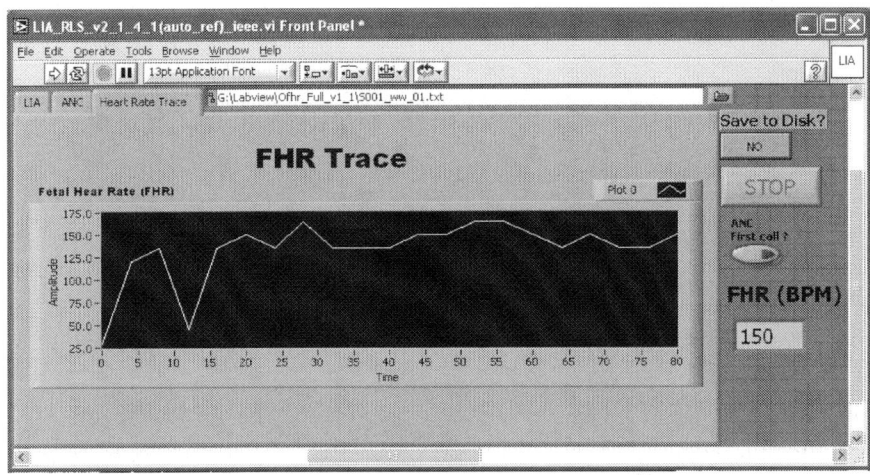

Figure 8. Graphical User Interface of OFHR system. FHR trace menu

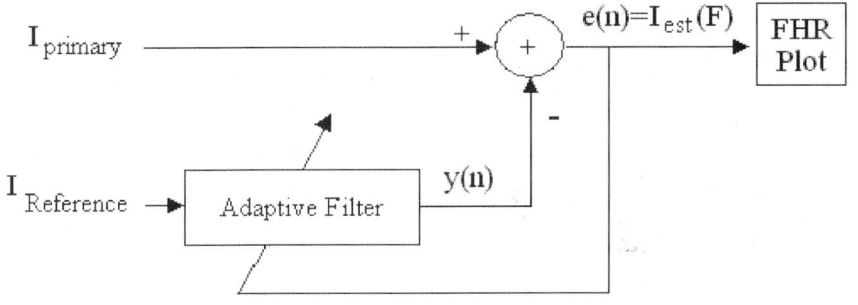

Figure 9. ANC block diagram

In this study, all fetuses were singleton with gestation weeks from 30 week to 40 week. Subjects with twin pregnancies, anterior placed placenta, obesity (BMI>30), gestational diabetes mellitus (GDM) and hypertension were excluded from this study. In addition, all fetuses in this study were found to be healthy by the obstetrician and born naturally (vaginally) without any complication.

During the data acquisition, the fetal probe is fixed to a maternal abdomen and the reference probe on her index finger in semi-Fowler position. The data analysis shows a correlation coefficient of 0.97 (p-value < 0.001) between optical and ultrasound FHR with a maximum error of 4%. Clinical results indicate that positioning the probe over the nearest fetal tissues (not restricted to head or buttocks) improves signal quality and therefore detection accuracy.

4. CONCLUSIONS

A low power OFHR detection system has been designed and developed using low cost, very low power (<68 mW) IR light and a commercially available silicon photo-detector. The digital synchronous detection and adaptive filtering techniques have been successfully implemented using LabVIEW 7.1. By applying digital synchronous detection and adaptive filtering techniques the FHR was determined with acceptable accuracy (maximum error of 4%) when compared to Doppler ultrasound. Attested by clinical results the probe positioning influences the acquired signal's quality and therefore affects the FHR results. Locating the nearest fetal tissues (not restricted to head or buttocks) to the probe will help to increase the signal quality and FHR determination accuracy.

The limitations of the optical technique are due to the presence of motion artifacts and sensitivity to the probe placement. The presence of motion artifacts may cause loss of correlation between the reference signal and the noise source (maternal PPG) in the mixed signal recorded from the maternal abdomen. The performance of the adaptive filtering scheme will suffer as a consequence, making the probe placement and stability an important criterion. Besides that, finding a proper location is needed in order to get signals with good SNR.

For the future development, by using an array of sensors to automatically select the channel with the highest SNR will eliminate the positioning problem. The topology of the sources and the photo-detector has to be determined. For the cost effective design, it is recommends that more light sources are used instead of photo-detectors. A wearable system will make the device more convenient for clinical applications in the near future. To ensure real-time and low-power function, the whole system can be implemented using embedded processor. The FHR will be wirelessly transmitted to another computing platform (PC or PDA) for further analysis, storage and transmission (to the nursing entity at a clinic). The main performance factors which will be considered are robustness, battery life, weight, dimensions and ergonomy. It is thought that the using of the selected platform (ARM) implementation will lead to a sufficiently low-cost bill-of material for the final product. During development phase, EMC directives will be taken into account so that the system's operation does not affect nor will be affected by other electronic devices. As a by-product of the project and contribution to the scientific community, it is proposed that acquired data during the project to be made available to a public data-base of biological signals (www.physionet.org) maintained by MIT in the USA.

NAN. ACKNOWLEDGEMENTS

This work has been partially supported by research university grant UKM-AP-TKP-07-2009. The authors would like to express their gratitude to Prof. Dr. M. A. J. M. Yassin and Associate Prof. Dr. S. Ahmad for their assistance in collecting the clinical data, and the staff at the Universiti Kebangsaan Malaysia Medical Centre, especially N. F. Mujamil for her assistance in determining the fetal position through ultrasound scan.

REFERENCES

1. S. V. Vaseghi, 2000 Advanced digital signal processing and noise reduction, Baffins Lane: John Wiley & Sons Ltd

2. B. Widrow, J. R. Glover Jr, J. M. Mc Cool, J. Kaunitz, C. S. Williams, R. H. Hearn, J. R. Zeidler, D. Eugene, R. C. Goodlin, 1975 Adaptive noise cancelling: principles and applications, Proceedings of the IEEE, 63 16921716

3. E. C. Ifeachor, B. W. Jervis, 2002 Digital signal processing: A practical approach, England: Prentice Hall

4. R. K. Freeman, T. J. Garite, M. P. Nageotte, 2003 Fetal heart rate monitoring, Lippincott William & Wilkins

5. J. S. Philip, 2002 Fetal distress. Current Obstetrics & Gynaecology, 12 1 521

6. R. Hershkovitz, E. Sheiner, M. Mazor, 2002 Ultrasound in obstetrics: a review of safety, European Journal of Obtetrics & Gynecology and Reproductive Biology, 101 1518

7. B. Karlsson, M. Berson, T. Helgason, R. T. Geirsson, L. Pourcelot, 2000 Effects of fetal and maternal breathing on the ultrasonic Doppler signal due to fetal heart movement, European Journal of Ultrasound, 4752

8. R. S. Khandpur, 2004 Biomedical Instrumentation: Technology and Applications, McGraw-Hill Professional

9. F. S. Najafabadi, E. Zahedi, Ali. M. A. Mohd, 2006 Fetal heart rate monitoring based on independent component analysis, Computers in Biology and Medicine, 36 3 241252

10. N. Ramanujam, G. Vishnoi, A. H. Hielscher, M. E. Rode, I. Forouzan, B. Chance, 2000 Photon migration through the fetal head in utero

using continuous wave, near infrared spectroscopy: clinical and experimental model studies, Journal of Biomedical Optics, 163172

11. H. Chance, 2005 Transabdominal Examination Monitoring and Imaging of Tissue. U.S. Patent 2005/0038344A1

12. A. Zourabian, B. Chance, N. Ramanujam, R. Martha, A. B. David, 2000 Trans-abdominal monitoring of fetal arterial blood oxygenation using pulse oximetry, Journal of Biomedical Optics, 5 391405

13. S. Nioka, M. Izzetoglu, T. Mawn, M. J. Nijland, D. A. Boas, B. Chance, 2005 Fetal transabdominal pulse oximeter studies using a hypoxic sheep model, The Journal of Maternal-Fetal and Neonatal Medicine, 17 6 393399

14. A. M. Vintzileos, S. Nioka, M. Lake, 2005 Transabdominal fetal pulse oximetry using near-infrared spectroscopy, American Journal of Obstetric & Gynaecology, 192 129133

15. E. Zahedi, G. K. Beng, 2008 Applicability of adaptive noise cancellation to fetal heart rate detection using photoplesthysmography. Computers in Biology and Medicine, 38 1 3141

16. R. Choe, T. Durduran, G. Yu, M. J. M. Nijland, B. Chance, A. G. Yodh, N. Ramanujam, 2003 Transabdominal near infrared oximetry of hypoxic stress in fetal sheep brain in utero, Proceedings of the National Academy of Sciences, 100 22 1295012954

17. K. B. Gan, E. Zahedi, Ali. M. A. Mohd, 2009 Trans-abdominal fetal heart rate detection using NIR photopleythysmography: instrumentation and clinical results, IEEE Transactions on Biomedical Engineering, 56 8 20752082 .

18. D. G. Manolakis, V. K. Ingle, S. M. Kogon, 2005 Statistical and adaptive signal processing. Norwood:Artech House, Inc.

19. S. Haykin, 2002 Adaptive filter theory. Prentice Hall.

20. K.B. Gan, 2009 Non-invasive fetal heart rate detection using near infrared and adaptive filtering. Available online from: (http://ptsldigital.ukm.my)

21. International Commission on Non-Ionizing Radiation Protection. 2000 ICNIRP statement on light-emitting diodes (LEDs) and laser

diodes: Implications for hazard assessment, Health Phys., 78 6 744752

22. J. D. Bronzino, 2000 The biomedical engineering handbook: 1 Florida: CRC Press LLC

23. J. Allen, A. Murray, 2000 Variability of photoplethysmography peripheral pulse measurements at the ears, thumbs and toes, IEEE Proceeding Science and Technology, 147 6 403407

Chapter 6

Noise Removal from EEG Signals in Polisomnographic Records Applying Adaptive Filters in Cascade

M. Agustina Garcés Correa and Eric Laciar Leber

[1] Gabinete de Tecnología Médica, Facultad de Ingeniería, Universidad Nacional de San Juan, Argentina

1. INTRODUCTION

Polisomnography (PSG) is the standard technique used to study the sleep dynamic and to identify sleep disorders. In order to obtain an integrated knowledge of different corporal functions during sleep, a PSG study must perform the acquisition of several biological signals during one or more nights in a sleep laboratory. The signals usually acquired in a PSG study include the electroencephalogram (EEG), the electrocardiogram (ECG), the electromiogram (EMG), the electro oculogram (EOG), the abdominal and thoracic breathings, the blood pressure, the oxygen saturation, the oro-nasal airflow and others biomedical records (Collop et. al., 2007).

Particularly, the EEG is usually analyzed by physicians in order to detect neural rhythms during sleep. However, it is generally contaminated with different noise sources and mixed with other biological signals. Their common artifacts sources are the power line interference (50 or 60 Hz), the ECG and EOG signals. Figure 1 shows an example of real EEG ECG and EOG signals recorded simultaneously in a PSG study. It can be seen that EEG signal is contaminated by the QRS cardiac complexes which appear as spikes at the same time in ECG record. Likewise, the low frequencies present in the contaminated EEG correspond to the opening, closing or movements of the eyes recorded in EOG signal. These noise sources increase the difficulty in analyzing the EEG and obtaining clinical information.

To correct, or remove the artifacts from the EEG signal many techniques have been developed in both, time and frequency domains (Delorme et. al., 2007; Sadasivan & Narayana, 1995). More recently, component-based techniques, such as principal component analysis (PCA) and independent component analysis (ICA); (Akhtar et. al., 2010; Astolfi et. al., 2010; Jung et. al., 2000), have also been proposed to remove the ocular artifacts from the EEG. The use of Blind Source Separation (BSS) (De Clercq et. al., 2005) and Parallel Factor Analysis (PFA) methods to remove artifacts from the EEG have been used in this area too (Cichocki & Amari, 2002; Makeig et. al., 2004). Wavelet Transform (WT) (Senthil Kumar et. al., 2009), WT combined with ICA (Ghandeharion et. al., 2009) and Autoregressive Moving Average Exogenous (ARMAX) (Hass et. al., 2003; Park et. al., 1998), have been applied too, to remove artifacts from EEG.

In this chapter, it is described a cascade of three adaptive filters based on a Least Mean Squares (LMS) algorithm to remove the common noise components present in the EEG signal recorded in polysomnographic studies.

Adaptive filters method has been used, among other applications, in external electroenterogram records (Mejia-García et. al., 2003) and in impedance cardiography (Pandey et. al., 2005). Other applications of this filtering technique in biomedical signals include, for example, removal of maternal ECG in fetal ECG records (Soria et. al., 1999) detection of ventricular fibrillation and tachycardia (Tompkins, 1993), cancellation of heart sound interference in tracheal sounds (Cortés, 2006), for pulse wave filter (Shen et. al., 2010), for tumor motion prediction (Huang et. al., 2010), detection of single sweep event related potential in EEG records (Decostre et. al., 2005), detection of SSVEP in EEG signals (Diez et. al., 2011) and for motor imagery (Jeyabalan et. al., 2007).

In the particular case of artifacts removal in EEG records, He et. al. (2007) studied the accuracy of adaptive filtering method quantitatively using simulated data and compared it with the accuracy of the domain regression for filtering ocular artifacts from EEG records. Their results show that the adaptive filtering method is more accurate in recovering the true EEG signals. Kumar et. al. (2009) shows that adaptive filtering can be applied to remove ocular artifacts from EEG with good results. Adaptive filters have been used to remove biological artifacts from EEG by others authors (Chan et. al., 1998; Karjalainen et. al., 1999; Kong et. al., 2001).

In order to improve the signal to noise ratio of EEG signals in PSG studies, we had proposed in a previous work a cascade of three adaptive filters based on a LMS algorithm (Garcés et. al., 2007). The first filter in the cascade eliminates line interference, the second adaptive filter removes the ECG complexes and the last one cancels EOG artifacts. Each stage uses a Finite Impulse Response (FIR) filter, which adjusts its coefficients to produce an output similar to the artifacts present in the EEG. In this chapter, we explain in detail the operation of the cascade of adaptive filters including novel tests to determinate the optimal order of FIR filter for each stage. Finally, we describe the results of the proposed filtering scheme in 18 real EEG records acquired in PSG studies.

2. MATERIALS

Eighteen PSG records belonging to sixteen subjects were selected from the MIT-BIH Polysomnographic Database. All subjects are aged 44 +/- 12 years. This database contains over 80 hours of four-, six-, and seven-channel PSG recordings. All of them contain EEG, ECG and Blood Pressure (BP) signals, some of them have Nasal or Plethysmograph Respiratory signals, five of them have O_2 Saturation signal, EOG and EMG signals. All the subjects have ECG signals annotated beat-by-beat, and EEG and respiration signals annotated by an expert with respect to sleep stages and apnea (Goldberger et. al., 2000). In this work were used only the EEG, ECG and EOG signals, all of them were sampled at 250 Hz.

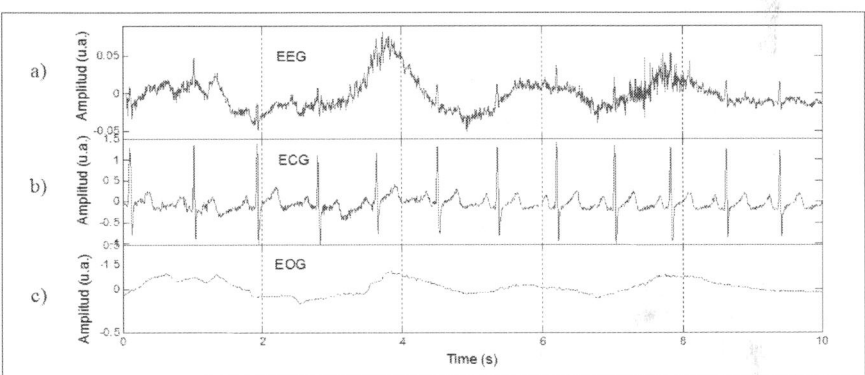

Figure 1. Some biological signals acquired in a PSG study a) EEG recording (corresponding to Patient 41) corrupted with ECG and EOG artifacts, b) Real ECG signal, and c) Real EOG signal.

3. COMMON ARTIFACTS IN EEG RECORDS

By artifacts it is understood all signals that appear in the EEG record which don't come from the brain. The most common artifacts in the EEG signal appear during the acquisition due to different causes, like as bad electrodes location, not clean hairy leather, electrodes impedance, etc. There is also a finding of physiological artifacts, that is, bioelectrical signals from other parts of the body (heart and muscle activity, eye blink and eyeball movement) that are registered in the EEG (Sörnmo & Laguna, 2005).

The problem of those artifacts is that they can made a mistake in the analysis of a EEG record, either in automatic method or in visual inspection by specialist (Wang et. al., 2000).

3.1. Power Line Interference

Biological records, especially EEG signals, are often contaminated with the 50 or 60 Hz line frequency interference from wires, light fluorescents and other equipments which are captured by the electrodes and acquisition system. The ignition of light of fluorescents usually causes artificial spikes in the EEG. They are distributed in several channels of EEG and can made a mistake in the analysis of the record (Sanei & Chambers, 2007)

3.2. Ocular Artifacts

The human eye generates an electrical dipole caused by a positive cornea and negative retina. Eye movements and blinks change the dipole causing an electrical signal known as an EOG. The shape of the EOG waveform depends on factors such as the direction of eye movements. A fraction of the EOG spreads across the scalp and it is superimposed on the EEG (Vigon et. al., 2000).

Two kinds of ocular artifacts can be observed in EEG records, eye blinks and eye movements. Eye blinks are represented by a low frequency signal (< 4 Hz) with high amplitude. It is a symmetrical activity mainly located on the front electrodes (FP1, FP2) with low propagation. Eye movements are also represented by a low frequency signal (< 4 Hz) but with higher propagation, (Crespel et. al., 2006). In order for the EEG to be interpreted for clinical use, those artifacts need to be removed or filtered from the EEG.

3.3. Cardiac Artifacts

Cardiac activity may have pronounced effects on the electroencephalogram (EEG) because of its relatively high electrical energy, especially upon the no-cephalic reference recordings of EEG. The QRS complexes appear in the EEG signal like regular spikes (Sörnmo & Laguna, 2005). In Figure 1 it can be observed the QRS complex present in a segment of EEG record. The QRS amplitudes in the ECG are of the order of mV, but in the external EEG they have been reduced. These artifacts in the EEG records could be clinically misleading.

3.4. Other Artifacts

The muscle disturbances are introduced in the EEG by involuntary muscle contractions of the patient, thus generating an electromyogram (EMG) signal present in the EEG record. The EMG and other biological artifacts have not been analyzing in the present work.

4. METHODOLOGY

Herein, we propose the use of adaptive filters to remove artifacts from EEG signal acquired in PSG studies. Usually, biological signals (ECG, EOG and others) have overlaped spectra with the EEG signal. For that, conventional filtering (band-pass, lower-pass or high-pass filters) cannot be applied to eliminate or attenuate the artifacts without losing significant frequency components of EEG signal.

Due to this reason, it is necessary to design specific filters to attenuate artifacts of EEG signals in PSG studies. The adaptive interference cancellation scheme is a very efficient method to solve the problem when signals and interferences have overlapping spectra.

Since the PSG recordings usually contain the ECG, EOG and EEG signals it is very convenient to apply this method to filter this kind of records.

4.1. Adaptive Filter

Adaptive filters are based on the optimization theory and they have the capability of modifying their properties according to selected features of the signals being analyzed (Haykin, 2005). Figure 2illustrates the structure of an adaptive filter. There is a primary signal $d(n)$ and a secondary signal $x(n)$.The linear filter $H(z)$ produces an output $y(n)$, which is subtracted from $d(n)$ to compute an error $e(n)$.

The objective of an adaptive filter is to change (adapt) the coefficients of the linear filter, and hence its frequency response, to generate a signal similar to the noise present in the signal to be filtered. The adaptive process involves minimization of a cost function, which is used to determine the filter coefficients. Initially, the adaptive filter adjusts its coefficients to minimize the squared error between its output and a primary signal. In stationary conditions, the filter should converge to the Wiener solution. Conversely, in non-stationary circumstances, the coefficients will change with time, according to the signal variation, thus converging to an optimum filter (Decostre & Arslan, 2005).

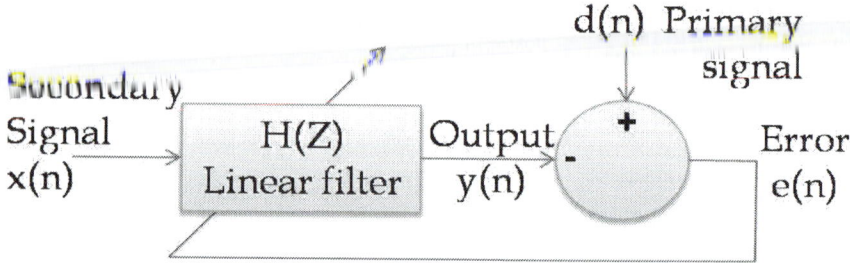

Figure 2. Structure of an adaptive filter.

In an adaptive filter, there are basically two processes:

- A filtering process, in which an output signal is the response of a digital filter. Usually, FIR filters are used in this process because they are linear, simple and stable.

- An adaptive process, in which the transfer function *H(z)* is adjusted according to an optimizing algorithm. The adaptation is directed by the error signal between the primary signal and the filter output. The most used optimizing criterion is the Least Mean Square (LMS) algorithm.

The structure of the FIR can be represented as,

$$y(n) = \sum_{k=0}^{L} w_k x(n-k) \tag{1}$$

where *L* is the order of the filter, *x(n)* is the secondary input signal, *wk* are the filter coefficients and *y(n)* is the filter output.

The error signal *e(n)* is defined as the difference between the primary signal *d(n)* and the filter output *y(n)*, that is,

$$e(n) = d(n) - y(n) \tag{2}$$

where,

$$e\left(n\right) = d\left(n\right) - \sum_{k=0}^{L} w_k x\left(n - k\right) \qquad (3)$$

The squared error is,

$$e^2\left(n\right) = d^2\left(n\right) - 2d\left(n\right)\sum_{k=0}^{L} w_k x\left(n - k\right) + \left[\sum_{k=0}^{L} w_k x\left(n - k\right)\right]^2 \qquad (4)$$

The squared error expectation for N samples is given by

$$\zeta = E\left[e^2\left(n\right)\right] = \sum_{k=0}^{N} e^2\left(n\right) \qquad (5)$$

$$\zeta = \sum_{n=1}^{N} \left[d^2\left(n\right)\right] - 2\sum_{k=0}^{L} w_k r_{dx}\left(n\right) + \sum_{k=0}^{L}\sum_{l=0}^{L} w_k w_l r_{xx}\left(k - l\right) \qquad (6)$$

where $r_{dx}(n)$ and $r_{xx}(n)$ are, respectively, the cross-correlation function between the primary and secondary input signals, and the autocorrelation function of the secondary input, that is

$$r_{dx}\left(n\right) = \sum_{n=1}^{N} d\left(n\right) x\left(n - k\right) \qquad (7)$$

$$r_{xx}\left(n\right) = \sum_{n=1}^{N} x\left(n\right) x\left(n - k\right) \qquad (8)$$

The objective of the adaptation process is to minimize the squared error, which describes a performance surface. To get this goal there are different optimization techniques. In this work, we used the method of steepest descent (Semmlow, 2004). With this, it is possible to calculate the filter coefficient vector for each iteration k having information about the previous coefficients and gradient, multiplied by a constant, that is,

$$w_k(n+1) = w_k(n) + \mu(-\nabla_k) \tag{9}$$

where μ is a coefficient that controls the rate of adaptation.

The gradient is defined as,

$$\nabla_k = \frac{\partial\{e^2(n)\}}{\partial w_k(n)} \tag{10}$$

Substituting (10) in (9) leads to,

$$w_k(n+1) = w_k(n) - \mu\frac{\partial\{e^2(n)\}}{\partial w_k(n)} \tag{11}$$

Deriving with respect to wk and replacing leads to,

$$w_k(n+1) = w_k(n) - 2\mu e(n)\frac{\partial\{e(n)\}}{\partial w_k(n)} \tag{12}$$

$$w_k(n+1) = w_k(n) - 2\mu e(n)\frac{\partial\left\{d(n) - \sum_{k=0}^{L} w_k x(n-k)\right\}}{\partial w_k(n)} \tag{13}$$

Since $d(n)$ and $x(n)$ are independent with respect to wk, then,

$$w_k(n+1) = w_k(n) - 2\mu e(n) x(n-k) \tag{14}$$

Equation (14) is the final description of the algorithm to compute the filter coefficients as function of the signal error $e(n)$ and the reference input signal $x(n)$. The coefficient μ is a constant that must be chosen for quick adaptation without losing stability. The filter is stable if μ satisfies the following condition, (Sanei & Chambers, 2007).

$$0 < \mu < \frac{1}{(10.L.\,P_{xx})}; P_{xx} \approx \frac{1}{M+1} \sum_{n=0}^{M-1} x^2(n) \qquad (15)$$

where L is the filter order and P_{xx} is the total power of the input signal.

4.2. Artifacts Removal From Eeg

As it is mentioned above, the adaptive interference cancellation is a very efficient method to solve the problem when signals and interferences have overlap spectra.

The adaptive noise canceller scheme is arranged on the basic structure showed in Figure 2, where the primary and secondary inputs are called as "corrupted signal" and "reference signal", respectively.

In this scheme, it is assumed that the corrupted signal $d(n)$ is composed of the desired $s(n)$ and noise $n0(n)$, which is additive and not correlated with $s(n)$. Likewise, it is supposed that the reference $x(n)$ is uncorrelated with $s(n)$ and correlated with $n0(n)$. The reference $x(n)$ feeds the filter to produce an output $y(n)$ that is a close estimate of $n0(n)$ (Tompkins, 1993).

To remove the main artifacts of the EEG signal, we propose a cascade of three adaptive filters (see Figure 3). The input $d_1(n)$ in the first stage is the EEG corrupted with artifacts (EEG + line-frequency + ECG + EOG). The reference $x_1(n)$ in the first stage is an artificial sine function generated with 50 Hz (or 60 Hz, depends on line frequency). The output of $H_1(z)$ is $y_1(n)$, which is an estimation of the line artifacts present in the EEG. This signal $y_1(n)$ is subtracted from the corrupted $d_1(n)$ to produce the error $e_1(n)$, which is the EEG without line-interference. The $e_1(n)$ error is forwarded as the corrupted input signal $d_2(n)$ to the second stage. The reference input $x_2(n)$ of the second stage can be either a real or artificial ECG. The output of $H_2(z)$ is $y_2(n)$, representing a good estimate of the ECG artifacts present in the EEG record. Signal $y_2(n)$ is subtracted from $d_2(n)$; its result produces error $e_2(n)$. Thus, we have obtained the EEG without line and ECG artifacts. Then, $e_2(n)$ enters into the third stage as the signal $d_3(n)$. The reference input $x_3(n)$ of filter $H_3(z)$ is also a real or artificial EOG and its output is $y_3(n)$, which is a replica of the EOG artifacts present in the EEG record. Such $y_3(n)$, subtracted from $d_3(n)$ gives error $e_3(n)$. It is the final output of the cascade filter, that is, the clean EEG without artifacts.

The reference signals ECG and EOG and the corrupted EEG were acquired simultaneously in polysomnographic studies. EEG, ECG and EOG records belonged to adult patients and were downloaded from the MIT-BIH Polysomnographic Databas-Physiobank (Goldberger et. al., 2000).

In section 4.3 there are present the tests that were carried out to determine the optimum order of $H_1(z)$, $H_2(z)$ and $H_3(z)$.

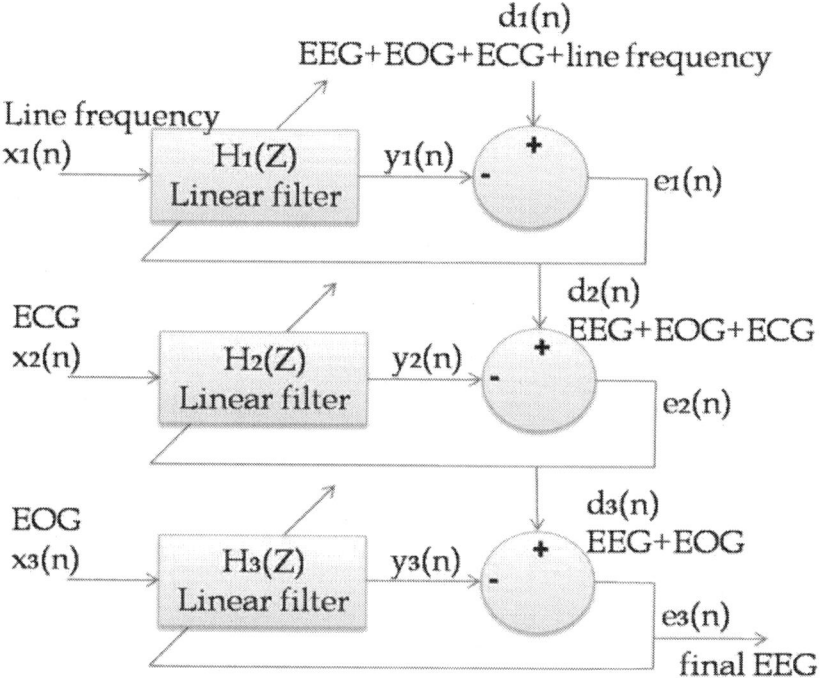

Figure 3. Structure of adaptive filters cascade for artifacts removal on EEG signal acquired in PSG studies.

4.3. Optimal Order of Fir Filters

To determine the optimum values of the orders L_1, L_2 and L_3 of $H_1(z)$, $H_2(z)$ and $H_3(z)$ filters the EEG signal were artificially contaminated with different coloured noises. The test to determinate the optimum values of the orders L_1, L_2 and L_3 was done with a coefficient convergence rates μ fixed in

0.001. As soon as the optimum value of the L of each stage was obtained the coefficient convergence rates μ of each stage was recalculated with Eq. (15) to assure an adequate adaptation. If μ is too big, the filter becomes unstable, and if it is too small, the adaptation may turn out too slow.

The tests were done using one stage of adaptive filter per time without using the cascade of three filters.

4.3.1. Optimal Estimation Of Order L_1 For Filter $H_1(Z)$.

The first stage filter attenuates the line frequency and was used to determinate the optimum value L_1 of $H_1(z)$. To determinate L_1, the EEG was artificially contaminated with a sinusoidal signal of 50 Hz which amplitude is adjusted in 30%, 50%, 80% and 100% of the Root Mean Square (RMS) value of original EEG signal. Then, the filter order L_1 was adjusted with different values of 8, 16, 32, 64 and 128.

In order to study the filter performance, we estimated the Power Spectral Density (PSD) of the original real EEG signal, the contaminated EEG and the different filtered versions of the EEG signal. PSD was computed using the Burg method with a model order equal to 12. Those graphics for one patient are presented in Figure 4 as an example.

Then, we estimated the normalized area below the frequency coherence function and the maximum of temporal cross-correlation normalized function between the filtered EEG signals and the contaminated EEG. If the signals are identical these parameters must be equal to 1. This test was done for each patient.

Table 1 show the averaged values of two parameters for all EEG records of the database.

Figure 4 is an example of PSD graphics for a EEG recording (corresponding to Patient 48) but all records of the database have a similar behaviour in the test. In this Figure it could be observed that as L_1 increases, the attenuation of the 50 Hz interference is more significant. However, if L_1 is higher than 32, it can be seen than other frequencies of spectrum are modified.

For this reason, there is a tradeoff between the 50 Hz interference attenuation and the modification of the main frequency components of EEG signal.

In table 1 it can be observed that the best option between $L_1=8$, $L_1=16$ and $L_1=32$ is $L_1=16$, because it have the minimum area of coherence and similar values of maximum in cross- correlation with $L_1=32$. Chosen this value of the order L_1 there is a loss of information of original signal and there is not a modification in the rest of the spectrum.

Table 1. Average values of the normalized parameters between filtered EEG signal and contaminated EEG signal with line interference for different values of L_1.

Contamination of line frequency	L_1	Coherence	Cross-correlation
30%	8	0.9943	0.976
	16	0.994	0.9727
	32	0.9947	0.9657
	64	0.9939	0.9497
	128	0.9912	0.9062
50%	8	0.9936	0.9426
	16	0.9932	0.9393
	32	0.9938	0.9326
	64	0.993	0.9171
	128	0.9902	0.8751
80%	8	0.9918	0.8739
	16	0.9914	0.8706
	32	0.9919	0.8643
	64	0.9909	0.85
	128	0.9879	0.8111
100%	8	0.9903	0.8223
	16	0.9898	0.8191
	32	0.9901	0.8131
	64	0.989	0.7996
	128	0.9859	0.7631

It is concluded that the optimum value of L_1 for the first filter is $L_1=16$ (for a sampling frequency of 250 Hz). For this order, the optimum value of the coefficient convergence rates μ calculated with Eq. (15) must be positive and lower than 0.047.

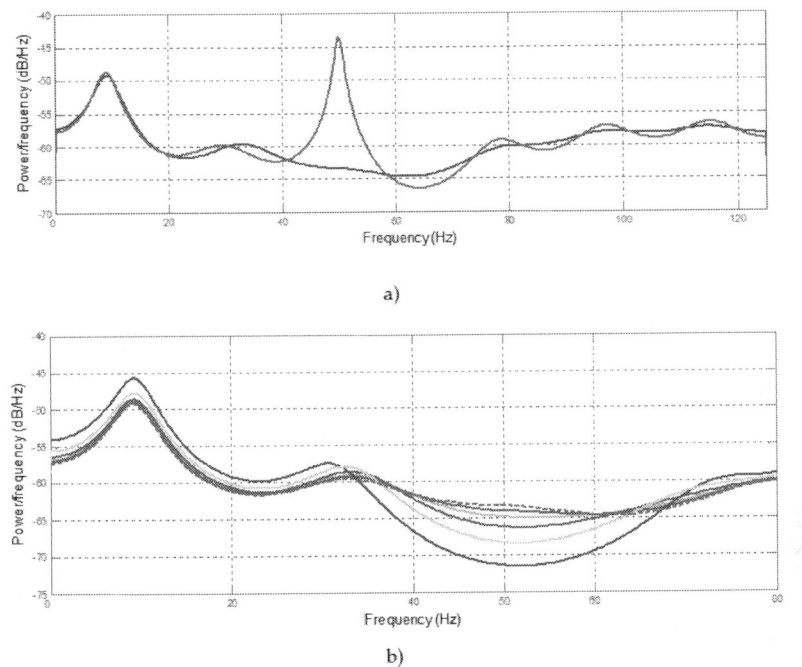

a)

b)

Figure 4. Power Spectral Density (PSD) of a EEG signal before and after the first adaptive filter $H1(z)$. a) In blue: PSD of original EEG, in red: PSD of EEG signal contaminated with an artificial line interference. b) PSD of filtered EEG signal for different values of the order $L1$. Red: original EEG, Green: $L1=8$, Orange: $L1=16$, Purple: $L1=32$, Light Blue: $L1=64$, Blue: $L1=128$.

4.3.2. Optimal Estimation Of Order L_2 For Filter $H_2(Z)$.

The second stage filter attenuates ECG artifacts (mainly QRS complexes) present in EEG signal, and was used to determinate the optimum value of the order L_2 of $H_2(z)$. To determinate L_2, the EEG was artificially contaminated with a coloured noise, with a -3dB bandwidth between 5 Hz and 40 Hz. This bandwidth was selected considering that QRS complexes have almost their total energy in this frequency band (Thakor, 1984). Then, the filter order L_2 was adjusted with the different values of 16, 32, 64, 128, 256 and 512.

As a similar way to optimum value estimation of L_1, we estimated the PSD of the original real EEG signal, the contaminated EEG and the different filtered versions of the EEG signal.

Figure 5 shows the PSD graphics for an EEG recording before and after the second adaptive filter. In this Figure it could be observed that the possible optimum values of L_2 in filter the cardiac frequencies between 5Hz and 40Hz are L_2=16, L_2=32 or L_2=64, because the rest of the values of L_2 modify the frequencies of the entire spectrum.

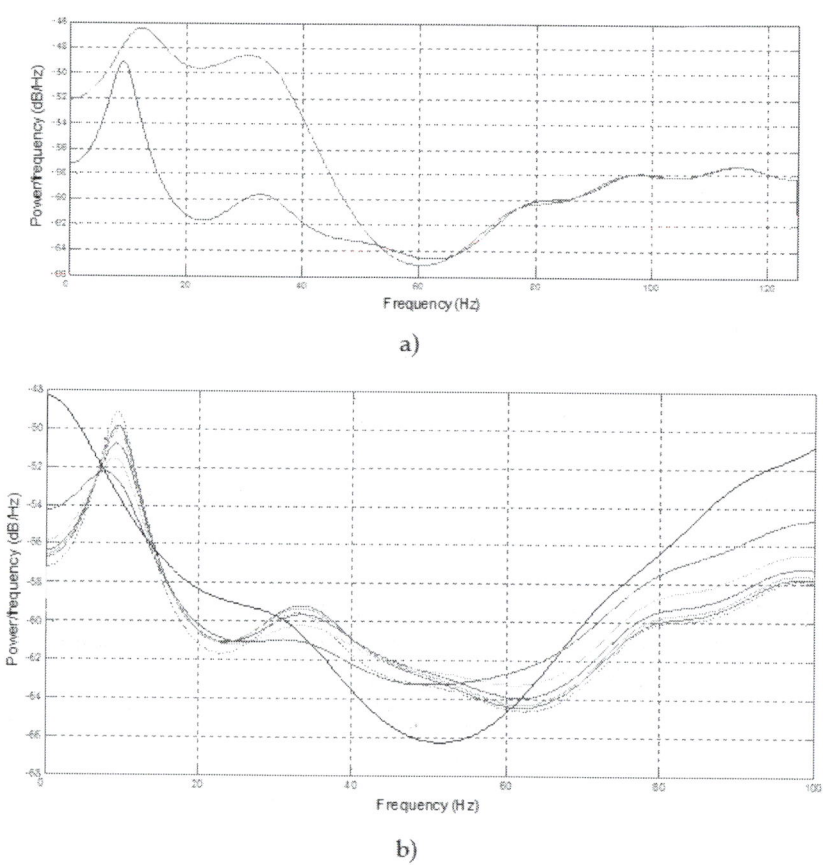

Figure 5. Power Spectral Density (PSD) of a EEG signal before and after the second adaptive filter $H2(z)$. a) In blue: PSD of original EEG, in red: PSD of EEG signal contaminated with coloured noise (5Hz to 40 Hz). b) PSD of filtered EEG for different values of the order $L2$. Red: original EEG, Green: $L2$=16, Orange: $L2$=32, Purple: $L2$=64, Light Blue: $L2$=128, Blue: $L2$=256, Black: $L2$=512.

Table 2 shows the average of the normalized area below the frequency coherence function and the maximum of temporal cross-correlation normalized function (between the filtered EEG signals and the contaminated EEG) for all recordings analyzed and for different values of $L2$.

TABLE 2. Average values of the normalized parameters between filtered EEG signal and contaminated EEG signal for different values of $L2$.

L2	Coherence	Cross-correlation
16	0.2588	0.5686
32	0.2596	0.5595
64	0.2927	0.5406
128	0.2641	0.5087
256	0.1756	0.4576
512	0.1579	0.3463

In table 2 it can be observed that the best option between $L_2=16$, $L_2=32$ or $L_2=64$ is $L_2=32$, because it have the minimum value of the normalized area below the frequency coherence function and the lower values of maximum of cross- correlation normalized function without losing information and not modifying the spectrum of the original EEG signal.

It is concluded that the optimum value of L_2 for second filter is $L_2=32$. For this order, the optimum value of the coefficient convergence rates μ calculated with (15) must be positive and lower than 0.02367.

4.3.3. Optimal Estimation of Order L_3 For Filter $H_3(Z)$.

As it is mentioned above, the third and last stage filter attenuates EOG artifacts present in EEG. In this section, we determinate the optimum value of the order L_3 of $H_3(z)$. To determinate it, the EEG was artificially contaminated with a coloured noise with a -3dB bandwidth between 0.5 Hz and 10 Hz. This bandwidth includes the main frequency components of EOG artifacts. Then, we evaluated the filter performance with different L_3 values (4, 8, 16 and 32).

As a similar way to optimum value estimation of L_1 and L_2, we estimated the PSD of the original real EEG signal, the contaminated EEG and the different filtered versions of the EEG signal.

Figure s 6 and 7 show the PSD graphics for an EEG recording before and after the third adaptive filter. It can be observed that all the values of the order L_3 chosen have good result to filter the frequencies lower than 10 Hz (see Figure 6). No one introduce interferences in other frequencies. But with values bigger than 256 it could be observed a distortion in high frequencies and a loss of information of the original signal in low frequencies (see Figure 7). The modification of the high frequencies and the losing of information in low frequencies are shown in Figure 7, where there have been filtered the contaminated EEG with values of $L_3=256$ and $L_3=512$

Table 3. Average values of the normalized parameters between filtered EEG signal and contaminated EEG signal for different values of *L3*

L3	Coherence	Cross-correlation
4	0.8773	0.8014
8	0.8586	0.7979
16	0.8579	0.7937
32	0.8584	0.7863
64	0.8586	0.7842

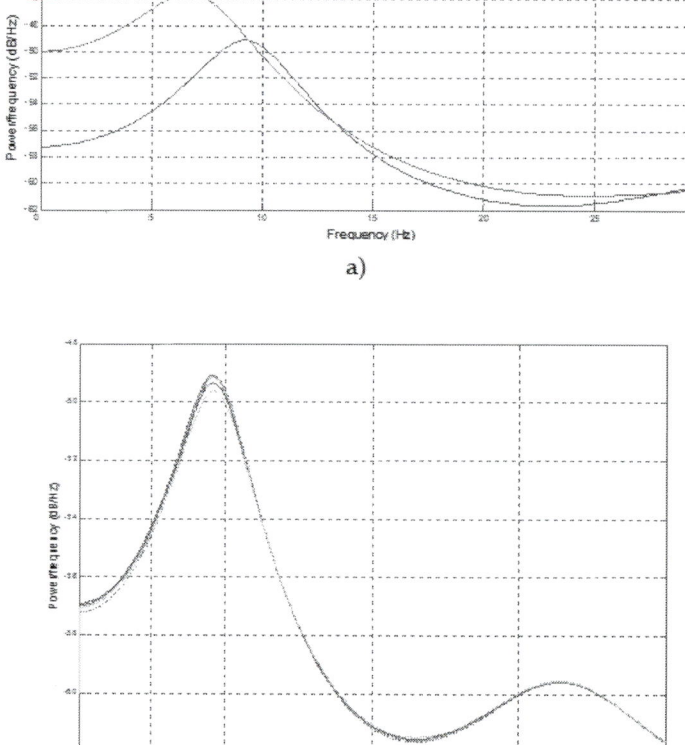

a)

b)

Figure 6. Power Spectral Density (PSD) of a EEG signal before and after the third adaptive filter $H_3(z)$. a) In blue: PSD of original EEG, in red: PSD of EEG signal contaminated with coloured noise (0.5 Hz to 10 Hz). b) PSD of EEG signal filtered for different values of the order L_3, Red: original EEG, Green:L_3=4, Orange: L_3=8, Purple: L_3=16, Light Blue: L_3=32,

Table 3 shows the averaged values of the normalized area below the frequency coherence function and temporal cross-correlation normalized function (between the filtered EEG signals and the contaminated EEG) for all recordings analyzed and for different values of L_3.

In Table 3 it can be observed that the best option of the value of the order L_3 for the third filter is $L_3=16$, because it have the minimum value of the normalized area below the frequency coherence function and the lower values of maximum of cross- correlation normalized function without losing information of original signal and not modifying the spectrum of the original EEG. The results of the test using values of L_3 bigger than $L_3=256$ have not been included in Table 3.

It is concluded that the optimum value of L_3 for the third filter is $L_3=32$. For this value, the optimum value of the coefficient convergence rates μ calculated with (15) must be positive and lower than 0.02367.

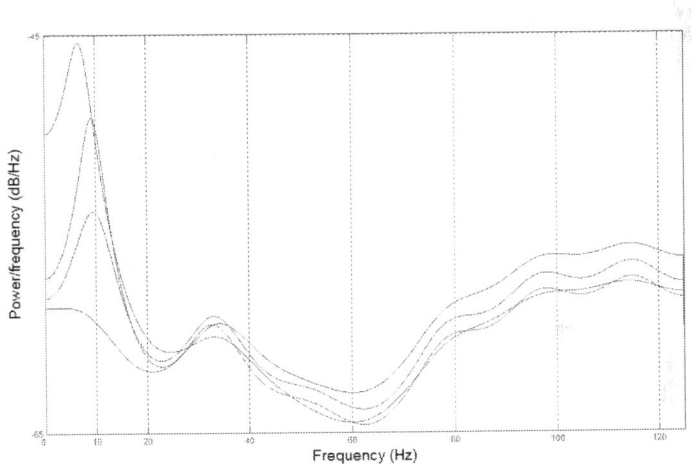

Figure 7. Power Spectral Density of a EEG signal before and after the third adaptive filter *H3(z)*. In Red: PSD of the original EEG signal. In Green: PSD of the EEG signal contaminated with coloured noise (0.5 Hz to 10 Hz). In Purple: PSD of the EEG filtered for order *L3=256*. In Blue: PSD of the EEG filtered for *L3=512*. Note the modification in high frequencies and losing of information in low frequencies.

5. RESULTS

Eighteen real EEG records acquired in PSG studies were processed with the cascade of adaptive filters. According to the previous tests, the values of the orders L_1, L_2 and L_3 were adjusted as L_1= 16, L_2= 32 and L_3= 32.

As it was mentioned in section 2, only five subjects from the entire database have EOG signals. So, the EEG signals of these five patients have been filtered with the entire cascade shown in Figure 3. The others thirteen EEG (belonging to the rest of the patients) have not been filtered with the last third stage.

The input $d_1(n)$ in the first stage is the EEG corrupted with artifacts (EEG + line-frequency + ECG + EOG). The reference $x_1(n)$ in the first stage is an artificial sine function generated with 50 Hz with the same RMS of the EEG signal. The $e_1(n)$, which is the EEG without line-interference, is forwarded as the corrupted input signal $d_2(n)$ to the second stage. The reference input $x_2(n)$ of the second stage is the real ECG. The error $e_2(n)$ is the EEG without line and ECG artifacts and enters into the third stage as the signal $d_3(n)$. The reference input $x_3(n)$ of filter $H_3(z)$ is a real EOG. The error $e_3(n)$ is the final output of the cascade filter, that is, the clean EEG without artifacts.

In order to study the filter performance we estimated the normalized area below the frequency coherence function and the maximum of temporal cross-correlation normalized function between the filtered EEG signals of each stage and the original EEG for the entire data base.

Table 4 shows the results obtained for each record of the database processed by the first stage of the propose filter. In this table, it is presented the values of the normalized area of frequency coherence function and the normalized maximum of temporal cross-correlation between the contaminated signal $d_1(n)$ and the error signal $e_1(n)$. Those values show that the first stage attenuates the line interference.

Figure 8 illustrates a temporal segment of 10s of the original EEG record (corresponding to Patient 41) and its filtered version after the first stage of adaptive filter. In this Figure it can be observed that the 50 Hz power line component is significantly filtered.

Figure 9 shows the PSD function of the same original and filtered EEG signals shown in Figure 8. The PSD of the filtered signal shows that the first stage attenuates the line-frequency artifacts. The $H_1(z)$filter adapts the amplitude and the phase of the artificial sinusoidal signal $x_1(n)$ (50Hz) in order to have as output a replica, $y_1(n)$, of the line-frequency artifacts present in the EEG.

After 50 Hz filtering, the EEG is forwarded to the second stage in order to remove ECG artifacts (seeFigure 3).

Table 4. Normalized area of frequency coherence function and maximum of temporal cross -correlation function between the signals $d_1(n)$ and $e_1(n)$ of the first stage of proposed filter.

Patient	Coherence %	Cross-correlation
1a	0.869	0.673
1b	0.8901	0.6349
2a	0.9833	0.4724
2b	0.9507	0.5417
3	0.9279	0.4044
4	0.9776	0.3615
14	0.9807	0.4698
16	0.9816	0.4452
32	0.9879	0.8309
37	0.9881	0.9293
41	0.9963	0.9857
45	0.9928	0.7017
48	0.9983	0.9413
59	0.9839	0.397
60	0.9747	0.2807
61	0.9663	0.4281
66	0.9783	0.4213
67	0.9734	0.5504
average	0.9667	0.5816

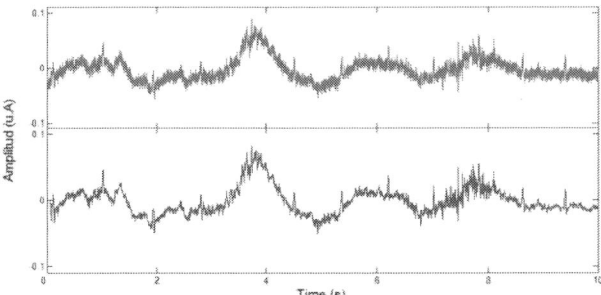

Figure 8. Example of a temporal segment of EEG filtered with stage 1 for patient 41. a) Red: Original EEG contaminated with 50 Hz power line interference, $d1(n)$.b) Blue: EEG without line interference, $e1(n)$.

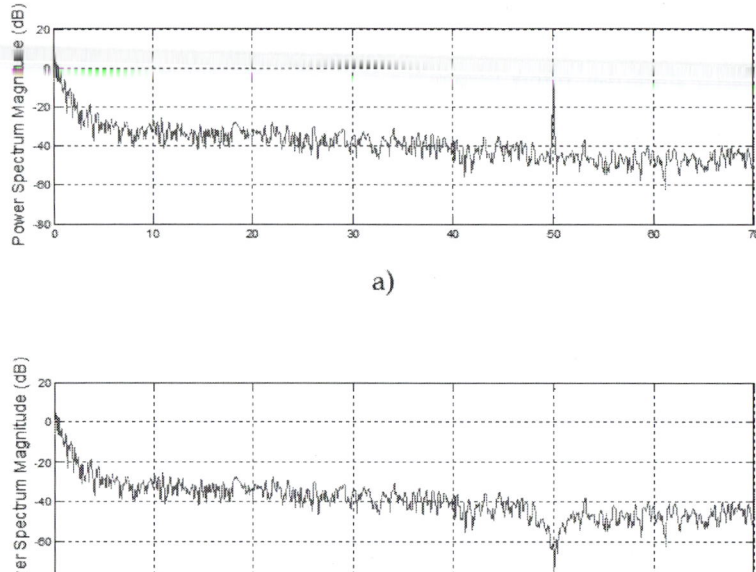

a)

b)

Figure 9. Example of first stage of the proposed filter. a) PSD of original EEG with artifacts. b) PSD of first stage output *e1(n)*, where the 50 Hz component is attenuated.

Table 5 shows the results obtained for each record of the database processed by the second stage. In this table, it is presented the values of the normalized area of frequency coherence function and the normalized maximum of temporal cross-correlation between the contaminated signal *d2(n)* and the error signal *e2(n)*. Those values show that the second stage attenuates QRS complexes artifacts introduced by ECG signal.

Figure 10 shows an example of 10s of EEG signal (corresponding to patient 41) processed by the second filter. The contaminated signal $d_2(n)$ is shown in red. It could be observed the presence and morphology similarity of QRS complexes of the ECG (in green) in the EEG record. The output signal $y_2(n)$ of $H_2(z)$ is drawn in black colour, this signal is an estimation of the ECG artifacts present in the EEG. The $H_2(z)$ filter adapts the amplitude and the phase of the reference signal $x_2(n)$ (ECG signal) in order to have as output a replica of the artifacts present in the EEG

After 50 Hz and ECG filtering, the EEG is forwarded to the third stage in order to remove EOG artifacts.

Table 5. Normalized area of frequency coherence function and maximum of temporal cross - correlation function between the signals $d_2(n)$ and $e_2(n)$ of the second stage of proposed filter.

Patient	Coherence	Cross-correlation
1a	0.8528	0.7514
1b	0.8801	0.518
2a	0.9709	0.9467
2b	0.9946	0.9845
3	0.9107	0.946
4	0.912	0.791
14	0.9276	0.8768
16	0.907	0.8757
32	0.8364	0.3333
37	0.855	0.6725
41	0.8204	0.7826
45	0.7985	0.7981
48	0.9096	0.6893
59	0.9106	0.5431
60	0.8224	0.3027
61	0.8979	0.2482
66	0.8097	0.5319
67	0.8464	0.8209
average	0.8342	0.6439

Table 6 shows the results obtained for five records of the database processed by the third stage. In this table, it is presented the values of the normalized area of frequency coherence function and the normalized maximum of temporal cross-correlation between the contaminated signal $d_3(n)$ and the error signal $e_3(n)$, which is the final output of the proposed filter. As it has been mentioned before only five patients have been filtered with the third stage, the rest of them do not have the reference signal $x_3(n)$. Those values show that this last stage attenuates artifacts introduced by the EOG.

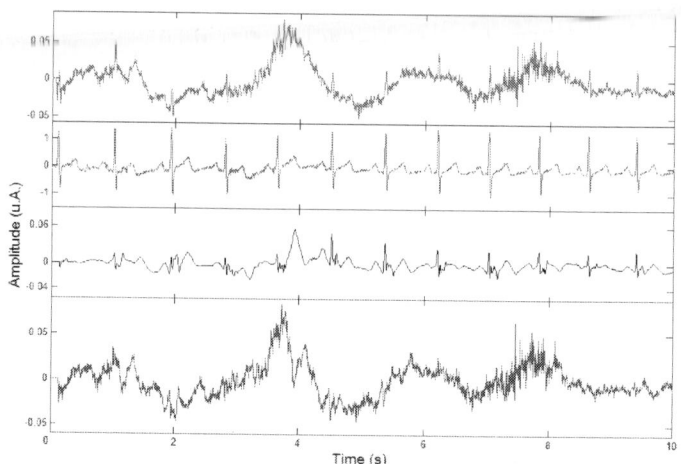

Figure 10. Example of a temporal segment of EEG filtered with stage 2 for patient 41. In Red: Contaminated EEG, $d_2(n)$. In Green: ECG signal. In Black: output signal from $H_2(z)$, that is $y_2(n)$. In Blue: EEG without ECG artifacts, $e_2(n)$.

Table 6. Normalized area of frequency coherence function and maximum of temporal cross - correlation function between the signals $d_3(n)$ and $e_3(n)$ of the third stage of proposed filter.

Patient	Coherence	Cross-correlation
32	0.9985	0.9907
37	0.9912	0.7949
41	0.9859	0.6052
45	0.999	0.95
48	0.9527	0.7943
average	0.9855	0.827

* Patient without available EOG signal.

Figure 11 shows the same 10s of temporal EEG signal of patient 41. There it can be observed all signals of third stage. The contaminated signal $d_3(n)$ is drawn in red colour. It can be observed the presence and morphology similarity of the EOG signal in the EEG record. The output signal $y_3(n)$ of $H_2(z)$ is in black colour in the Figure , this signal is an estimation of the EOG signal present in the EEG. The $H_3(z)$ filter adapts the amplitude and the phase of the reference signal $x_3(n)$ (EOG signal) in order to have as output a replica of the EOG artifacts present in the EEG.

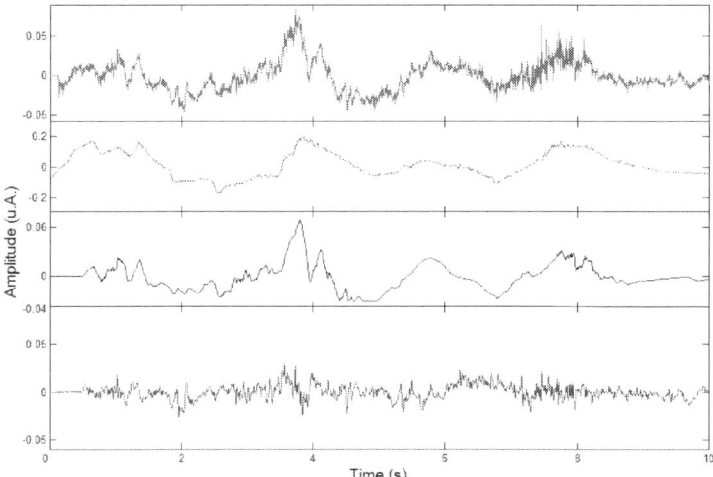

Figure 11. Example of temporal segment of EEG filtered with stage 3 for patient 41. In Red: Contaminated EEG, $d_3(n)$. In Green: EOG signal. In Black: output signal from $H_3(z)$, that is $y_3(n)$. In Blue: EEG without EOG artifacts, $e_3(n)$.

Figure 12 show the PSD of the contaminated EEG of third stage, $d_3(n)$, of the reference signal $x_3(n)$, EOG, and of the filtered EEG signals illustrated in Fig. 11. Note that the low frequencies of the EOG present in the contaminated EEG are attenuated in the filtered EEG signal.

Figure 13 is shown temporal temporal segments of 10s of EEG. In this Figure it could be observed the attenuation of line frequency and biological artifacts without losing important information of the EEG signal. Results show that the proposed adaptive filter cancels correctly the line frequency interference and attenuate very well the biological artifacts introduced by the ECG and the EOG.

6. DISCUSSION AND CONCLUSION

In this chapter, a novel filtering method based on three adaptive filters in cascade has been proposed to cancel common artifacts (line interference, ECG and EOG) present in EEG signals recorded in PSG studies.

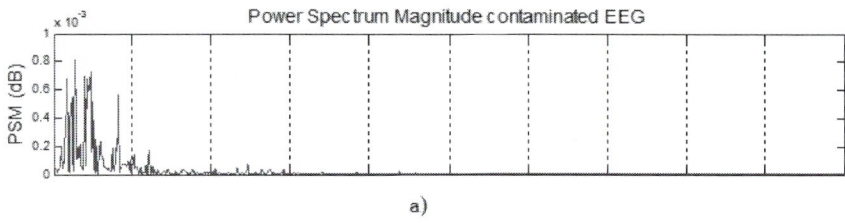

a)

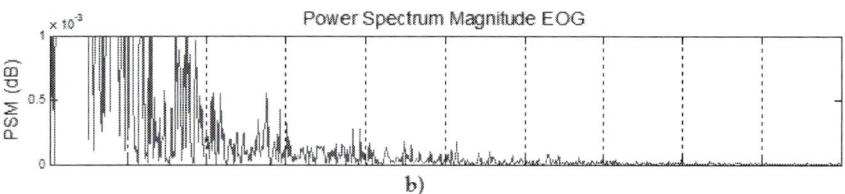

b)

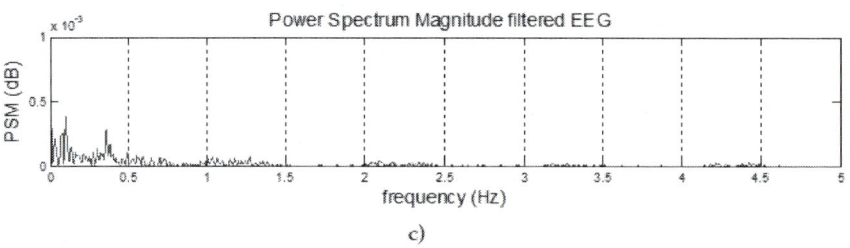

c)

Figure 12. Example of third stage of the proposed filter a) PSD of the contaminated EEG, $d_3(n)$, b) PSD of the reference signal $x_3(n)$, EOG, c) PSD of the filtered EEG signal.

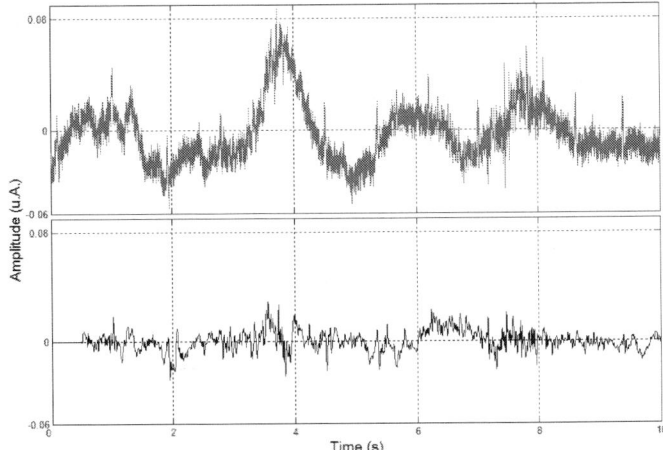

Figure 13. Example of temporal segments of contaminated EEG and EEG filtered with the entire cascade for patient 41. In Red: Contaminated EEG, $d_1(n)$, In Black: final filtered EEG without line interference, ECG and EOG artifacts, $e_3(n)$.

Other methods (like PCA, ICA, BSS or WT) have been described in the bibliography to cancel these artifacts in the EEG signals. However, those methods have some restrictions. For example, the properties of WT make it has an advantage in processing short-time instantaneous signal, but it needs that the frequency range of the EEG signal was not overlap with the bandwidth of noise sources and in this case the frequencies bands of the ECG and EOG signal are overlap with the frequencies of the EEG. ICA is a developed method for transforming an observed multidimensional vector into components that are statistically as independent from each other as possible. This method needs that the dimension of the signals were larger than that of original signals, and every original signal must be non-Gaussian. With more observed signals ICA will get better filtering result, which limits the application of this technique in few channels EEG recordings.

The main advantages of the proposed adaptive filtering method can be summarized as:

- The method does not have restrictions about the signal to be filtered.
- The implementation of adaptive filtering is very simple and fast and the results can be obtained without complex calculations.
- The filter coefficients can be adapted to variations in heart frequency, abrupt changes in the line frequency (caused, say, by ignition of electric devices) or modifications due to eye movements.

- At each stage output, the error signals $ei(n)$, EEG with one of the three attenuated artifacts are present; such separation (by artifacts) may be useful in some applications where such output might be enough.
- The filters have a linear phase response so no phase distortion is made. This is particularly important for the analysis of neurological rhythms in EEG signals

As soon as the optimal orders of the three filters were determinate, the method was tested in 18 real EEG records acquired in PSG studies. Figure 13 is a good example of an EEG record corrupted by three types of artifacts and its corresponding filtered version. It can be seen that all artifacts have been eliminated or attenuated, improving the quality of EEG record. The remaining records analyzed in the work had obtained similar results and their filtered EEGs don't have large artifacts.

It has been concluded that proposed adaptive filtering scheme with the appropriate values of order L_i, attenuate correctly ECG, EOG and line interference without removing significant information embedded in EEG signals registered in PSG studies. Due to the fact that the these studies usually have the ECG, EOG and EEG signals, the proposed cascade of adaptive filters is very useful and appropriate for the analysis of PSG recordings in sleep laboratories. The cascade could be used in others biomedical applications and in BCI applications.

NAN. ACKNOWLEDGEMENTS

This work has been supported by grants from Consejo Nacional de Investigaciones Científicas y Técnicas (CONICET) and Universidad Nacional de San Juan (UNSJ), both Argentinian institutions.

REFERENCES

1. M. T. Akhtar, C. J. James, W. Mitsuhashi, 2010 Modifying the Spatially-Constrained ICA for Efficient Removal of Artifacts from EEG Data. Proceedings of 4th International Conference on Bioinformatics and Biomedical Engineering (iCBBE), 14 . 978-1-42444-713-8 Chengdu, China, June 18-20, 2010.

2. L. Astolfi, F. Cincotti, D. Mattia, F. Babiloni, M. G. Marciani, Fallani. F. De Vico, M. Mattiocco, F. Miwakeichi, Y. Yamaguchi, P. Martinez, S. Salinari, A. Tocci, H. Bakardjian, F. B. Vialatte, A. Cichocki, 2006 Removal of ocular artifacts for high resolution EEG studies: a simulation study, Proceedings of 28th Annual International

Conference of the IEEE Engineering in Medicine and Biology Society, 976979 , 1-42440-033-3 York City, USA, Aug 30-Sept 3, 2006.

3. F. H. Y. Chan, W. Qiu, F. K. lam, P. W. F. Poon, 1998 Evoke potential estimation using modified time-sequence adaptive filter, Medical & Biological Engineering & Computing, 36 4 (July 1998) 407414

4. A. Cichocki, S. Amari, 2002 Adaptive Blind Signal and Image Processing, John Wiley & Sons. 0-47160-791-6

5. N. A. Collop, W. M. Anderson, B. Boehlecke, D. Claman, R. Goldberg, D. J. Gottlieb, 2007 Clinical guidelines for the use of unattended portable monitors in the diagnosis of obstructive sleep apnea in adult patients. Portable Monitoring Task Force of the American Academy of Sleep Medicine. Journal Clinical Sleep Medicine. 3 7 (December 2007), 737747 .

6. S. Cortés, R. Jane, A. Torres, J. A. Fiz, J. Morera, 2006 Detection and Adaptive Cancellation of Heart Sound Interference inTracheal Sounds, Proceedings of the 28th IEEE EMBS Annual International Conference,28602863 , 1-42440-033-3 York City, USA, Aug 30-Sept 3, 2006.

7. A. Crespel, P. Gélisse, M. Bureau, P. Genton, 2005 Atlas of Electroencephalography (1st ed), 1 2, J. Libbey Eurotext, 2-74200-600-1

8. W. De Clercq, A. Vergult, B. Vanrumste, J. Van Hees, A. Palmini, W. Van Paesschen, S. Van Huffel, 2005 A new muscle artifact removal technique to improve the interpretation of the ictal scalp electroencephalogram, Proceedings of the 27th Annual Conference IEEE Engineering in Medicine and Biology, 944947 ,0-78038-741-4 China, September 1-4, 2005.

9. A. Decostre, B. Arslan, 2005 An Adaptive Filtering Approach to the Processing of Single Sweep Event Related Potentials Data, Proceedings 5th International. Workshop Biosignal Interpretation, 13 , Tokyo, Japan, September 6-8, 2010.

10. A. Decostre, B. Arslan, 2005 An Adaptive Filtering Approach to the Processing of Single Sweep Event Related Potentials Data, Proceeding 5th International. Workshop Biosignal Interpretation, 13 ,Tokyo- Japan, September 2005

11. A. Delorme, T. Sejnowski, S. Makeig, 2007 Enhanced detection of artifacts in EEG data using higher order statistics and Independent component analysis, NeuroImage, 34 (2007), 14431449 , 1053-8119

12. P. Diez, Correa. M. A. Garcés, E. Laciar, 2010 SSVEP Detection using Adaptive filters, V Congreso Latinoamericano de Ingeniería Biomédica (CLAIB2011), Habana, Cuba, May 16-21, 2011, In press.

13. Correa. A. Garcés, E. Laciar, H. D. Patiño, M. E. Valentinuzzi, 2007 Artifact removal from EEG signals using adaptive filters in cascade, Journal of Physics, 90 (September 2007), 110 ,

14. H. Ghandeharion, H. Ahmadi-Noubari, 2009 Detection and Removal of Ocular Artifacts using Independent Component Analysis and Wavelets, Proceedings of the 4th International IEEE EMBS Conference on Neural Engineering, 653656 , 978-1-42442-073-5 Antalya, Turkey, April 29- May 2, 2009.

15. A. L. Goldberger, L. A. Amaral, N. L. Glass, J. M. Hausdorff, P. C. Ivanov, R. G. Mark, J. E. Mietus, G. B. Moody, C. K. Peng, H. E. Stanley, 2000 PhysioBank, PhysioToolkit, andPhysioNet: Components of a New Research Resource for Complex Physiologic Signals, Circulation, 101 23 215220 , June 2000

16. S. H. Hass, M. G. Frei, I. Osorio, B. Pasik-Duncan, J. Radel, 2003 EEG ocular artifact removal through ARMAX model system identification using extended least squares, Comunication in formation and system, 3 1 (June 2003), 1940 .

17. S. Haykin, 2005 Neural Network (2nd), Pearson Prentice Hall, 8-17808-300-0 India.

18. P. He, M. Kahle, G. Wilson, C. Russell, 2005 Removal of Ocular Artifacts from EEG: A Comparison of Adaptive Filtering Method and Regression Method Using Simulated Data, Proceedings of the 2005 IEEE Engineering in Medicine and Biology 27th Annual Conference, 11101113 , 0-78038-740-6 China, September 1-4, 2005

19. K. Huang, I. Buzurovic, Y. Yu, T. K. Podder, 2010 A Comparative Study of a Novel AE-nLMS Filter and Two Traditional Filters inPredicting Respiration Induced Motion of the Tumor, 2010 IEEE International Conference on Bioinformatics and Bioengineering, 281282 , 978-0-76954-083-2 Philadelphia, May 31-Jun3, 2010.

20. V. Jeyabalan, A. Samraj, L. Chu-Kiong, 2007 Motor Imaginary Signal Classification Using Adaptive Recursive Bandpass Filter and Adaptive Autoregressive Models for Brain Machine Interface Designs International Journal of Biological and Life Sciences, 3 2 March 2007), 116123 . e2010-3832

21. T. P. Jung, S. Makeig, M. Westerfield, J. Townsend, E. Courchesne, T. J. Sejnowski, 2000 Removal of eye activity artifacts from visual event-related potentials in normal and clinical subjects. Clinical Neurophysiology, 111 10 (October 2000), 17451758 . 1388-2457

22. P. Karjalainen, J. Kaipio, A. Koistinen, M. Vauhkonen, 1999 Subspace regularization method for the single trial estimation of evoked potentials, IEEE Transactions on Biomedical Engineering, 46 7 (July 1999), 849860 , 00189294

23. X. Kong, T. Qiu, 2001 Latency change estimation for evoked potentials a comparison of algorithms, Medical & Biological Engineering & Computing, 39 2 (2001), 208224 .

24. S. Makeig, S. Debener, J. Onton, A. Delorme, 2004 A. Mining event-related brain dynamics. Trends Cogn Sci., 204210 ., May 2004; 8 5

25. J. H. Mejia-Garcia, J. L. Martinez-Juan De, J. Saiz, J. Garcia-Casado, J. L. Ponce, 2003 Adaptive cancellation of the ECG interference in external electroenterogram, Proceedings of the 25th Annual International Conference of the IEEE Engineering in Medicine and Biology Society 2003,3 26392642 , 0-78037-789-3 Mexico September 17-21, 2003.

26. V. K. Pandey, P. C. Pandey, 2005 Cancellation of Respiratory Artifact in Impedance Cardiography, Proceedings of the 2005 IEEE Engineering in Medicine and Biology 27th Annual Conference, 55035506 , 0-78038-740-6 China, September 1-4, 2005

27. H. Park, J. , D. Jeong, U. , K.-S. Park, 2002 Automated Detection and Elimination of Periodic ECG Artifacts in EEG Using the Energy Interval Histogram Method. IEEE transactions on Biomedical Engineering. 49 12 (December 2002), 15261533 , 0018-9294

28. P. K. Sadasivan, D. Narayana, 1995 Line interference cancellation from corrupted EEG signals using Modified linear phase FIR digital filters, Engineering in Medicine and Biology Society, 1995 and 14th

Conference of the Biomedical Engineering Society of India. An International Meeting, Proceedings of the First Regional Conference, 3 35-3.38. India, February 15-18, 1995.

29. S. Sanei, J. Chambers, 2007 EEG Signal Processing (1st ed.), Jhon Wiley & Sons, 139780470025819, England.

30. J. L. Semmlow, 2004 Biosignal and Medical Image Processing, Marcel Dekker, 0-82474-803-4 State of America.

31. Kumar. P. Senthil, R. Arumuganathan, C. Vimal, 2009 An Adaptive method to remove ocular artifacts from EEG signals using Wavelet Transform, Journal of Applied Sciences Research, 5 7 (2009), 741745 , 741745

32. Z. Shen, C. Hu, M. Q. Meng, H. , 2010 A Pulse Wave Filter Method Based on Wavelet Transform Soft-threshold and Adaptive Algorithm, Proceedings of the 8th World Congress on Intelligent Control and Automation, 19471952 , 978-1-42446-712-9 Jinan, China, July 6-9, 2010.

33. E. Soria, M. Martínez, J. Calpe, J. F. Guerrero, A. J. Serrano, 1999 A new recursive algorithm for extracting the fetal ECG, Revista Brasileira de Engeñaría Biomédica, 15 (1999), 135139 . 15173151

34. L. Sörnmo, P. Laguna, 2005 Bioelectrical signal processing in cardiac and neurological applications (1st ed.), Elseiver, 0-12437-552-9

35. N. V. Thakor, J. G. Webster, W. J. Tompkins, 1984 Estimation of QRS complex power spectrum for design of a QRS filter, IEEE Transtaction Biomededical. Engenier., 31 11 (1984), 702706 .

36. W. J. Tompkins, 1993 Biomedical digital Signal Processing, Prentice Hall, New Jersey,

37. L. Vigon, M. R. Saatchi, J. E. W. Mayhew, R. Fernandes, 2000 Quantitative evaluation of techniques for ocular artifact filtering of EEG waveforms, IEE Processing Science Meas. Technology, 147 5 (September 2000), 219228 .

38. Y. L. Wang, J. H. Liu, Y. C. Liu, 2008 Automatic Removal of Ocular Artifacts from Electroencephalogram Using Hilbert-Huang Transform, The 2nd International Conference on Bioinformatics and Biomedical Engineering, ICBBE 2008, 21382141 , 978-1-42441-748-3 Shanghai, China, May 16-18, 2008.

Chapter 7

Adaptive Filtering by Non-Invasive Vital Signals Monitoring and Diseases Diagnosis

Omar Abdallah[1, 2], Armin Bolz[1] and S. Rajendran[1]

[1] *Institute for Biomedical Engineering, Karlsruhe Institute of Technology, Germany*
[2] *Biomedatronik, Karlsruhe, Germany*

1. INTRODUCTION

The reliability, reproducibility and accuracy of in-vivo measurements are of great importance and have to be thoroughly studied and to a great extend achieved. Reproducibility problems may result from the electronic components of the applied devices and the variability of measured variables as well as noise sources. The inaccuracy is caused by the approximation in the calculations or the used methods and by diverse sources of errors resulting from the subject under considerations and its surroundings. In sensible measurement like blood components, the positioning of the measuring sensor as well as the variation in the applied pressure and the characteristics of contact area between sensor and skin have a great effect on the accuracy and reproducibility of the measurements. The ambient noise like high frequency and line frequency (50 or 60 Hz) noise can be filtered by the detected biosignals like Photoplethysmogram (PPG) using the conventional analog or digital filters without a great effort. The motion artifact of the subject caused by him as well as by physical motion of body parts or by the surrounding has a varying frequency which may have the same range of the signal frequency. It is difficult to filter noise from these signals, and errors resulting from filtering can distort them. Usually physicians are misled by these noisy signals and the analysis can go wrong. An adaptive filter is essential by bio-signal and bio-image processing for noise cancellation without destroying or manipulating the valuable detected information.

Biomedical signals such as photoplethysmogram (PPG) (Figure 1), electrocardiogram (ECG), electroencephalogram (EEG), electromyogram (EMG) and impedance cardiogram (ICG) are very important in the diagnosis of different pathological variations. By the detection of these bio-signals as well as by the further derived parameters like oxygen saturation by pulse oximeter, the motion artifact is a great challenge, which may lead to erroneous results or even no results can be delivered [Lee].

The effectiveness of ECG monitors can be significantly impaired by motion artifact, which can cause misdiagnoses, lead to inappropriate treatment decisions or trigger false alarms. However, it is difficult to separate the noise from bio-signal due to its frequency spectrum overlapping that of the ECG. A portable ECG recorder using accelerometer based on motion artifact removal technique will be a great help for patients for tele-homecare or ambulatory ECG monitoring.

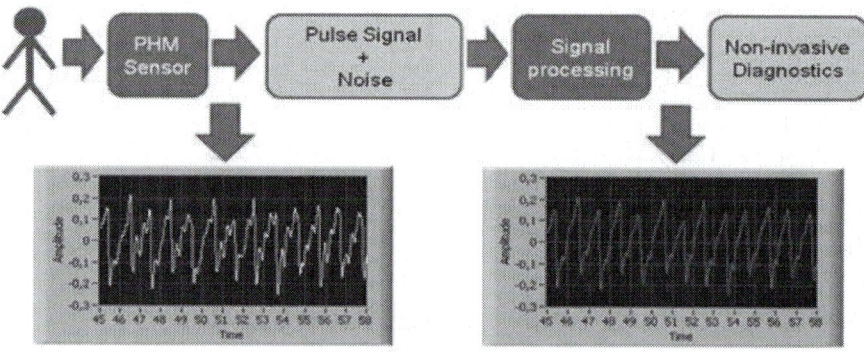

Figure 1. Signal detection and processing by noninvasive diagnosis

A maternal electrocardiogram (mECG) and abdominal noise in abdominal maternal recordings (especially by cardiotocography) can be orders of magnitude stronger than the fetal electrocardiogram (fECG) signal. An adaptive filter using frequency-domain or time-domain electrocardiogram features can be applied by the automatically extraction of a beat-to-beat fECG from mECG using surface electrodes placed on the maternal abdomen [Rik] [Prasad]. This will allow early diagnosis and monitoring treatment of certain fetal cardiac disorders.

By non-contact ECG monitoring, cardiac activity and movements (that may be seen in part in cardioballistogram CBG) may cause also disturbance to the detected signals, which can be eliminated by applying an appropriate adaptive filter

High-quality EEG recording is crucial for diagnosis of different pathological variations. EEG has biological artifacts and external artifacts. Biological artifacts can be EMG-, EOG- (Electrooculograph) CBG or ECG-signal [Rasheed]. These artifacts appear as noise in the recorded EEG signal individually or in a combined manner. These noise sources increase the difficulty in analyzing the EEG and to obtaining clinical information. For this reason, it is necessary to design specific filters to decrease such artifacts in EEG records. EEG quality in the MR scanner is compromised by artifacts caused by interaction between the subject, EEG electrode assemblies, and the scanner's magnetic fields [Rasheed 2009]. The three most significant causes of EEG artifacts in the scanner are the large movements in the static field like swallowing; the cardioballistogram and blood flow effects in the field associated with the subject's pulse; and the changing fields applied during fMRI image acquisition. Pulse artifact is potentially a significant problem as it is normally large amplitude, widespread on the scalp, and continuous. Using a cascade of adaptive filters based on a least mean squares (LMS) algorithm can eliminate the undesired signals or interferences.

2. PHOTOPLETHYSMOGRAM

The photoplethysmogram (PPG) waveform comprises a pulsatile physiological waveform superimposed on a slowly varying baseline with various lower frequency components. The pulsatile one is attributed to cardiac synchronous changes in the blood volume with each heart beat, and the second is attributed to respiration, sympathetic nervous system activity and thermoregulation. Figure 2 shows a typical PPG signal without motion artifact. The PPG technology has been used in a wide range of commercially available medical devices for measuring oxygen saturation, blood pressure and arterial stiffness, cardiac output, assessing autonomic function and detecting peripheral vascular diseases. Although the origins of the components of the PPG signal are not fully understood, there is no doubt that they can provide valuable information about the cardiovascular system and autonomic nervous system. Hence, there is a great interest in the technique in recent years, driven by the demand for low cost, very compact size, simple and portable technology for the primary care and community based clinical settings, non-invasive technology without side effects or risks as well as online monitoring capability and the advancement of computer-based pulse wave analysis techniques and diagnosis [Allen, Abicht]. A computer aided analysis tool for the hemodynamic diagnosis using PPG can be very helpful in clinical applications. Automatic assessment of the reliability of reference heart rates from patient vital-signs monitors incorporating both ECG and PPG based pulse measurements has been proposed by Yu et al. They expressed reliability as a quality index for each reference heart rate. The physiological waveforms were assessed using a support vector machine classifier and the independent computation of heart rate made by an adaptive peak identification technique that filtered out motion induced noise [Allen].

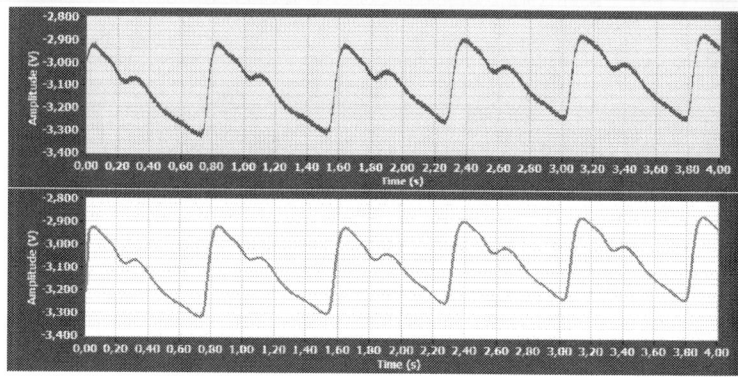

Figure 2. Photoplethysmogram PPG; top: detected raw signal, bottom: filtered signal

Also, due to demographic change, especially in the industrial countries, the personal health care of old people is of great importance for prevention and rehabilitation. Continuous monitoring of vital parameters is essential for that aim. By long term as well as by emergency, a monitoring without interruption is crucial for the diagnosis of the case under consideration. In many cases, a motion artifact caused by patient as well as by physical motion of body parts or by the surrounding may have the same range of the signal frequency. It is difficult to filter noise from these signals, and errors resulting from filtering can distort them. Pulse oximeter for measuring oxygen saturation (SpO_2) using more than one PPG signal is a valuable device for monitoring patients in critical conditions. PPG and the derived oxygen saturation are susceptible for motion artifact.

Pulse oximetry sensors use two Light Emitting Diodes (LEDs) which emit red and infrared light that shine through a reasonably translucent part of the patient's body. In pulse oximetry, it is called red light to the light band whose wavelength is comprised between 600-750 nm, while infrared light's wavelength varies between 850 and 1000 nm. These two wavelengths values are chosen because light absorption coefficient varies with the oxygen concentration of in both the red and the infrared light.Figure 3 shows the two principles of pulse oximetry: transmission and reflection pulse oximetry. By transmission Pulse oximetry, the light sensitive photodetector (Photodiode PD), which acts as a receiver picking up the light that passes through the measuring site, is opposite to the light emitter (light emitting diode LED). By reflection pulse oximetry the PD and the LED`s lie at the same side of the irradiated body portion.

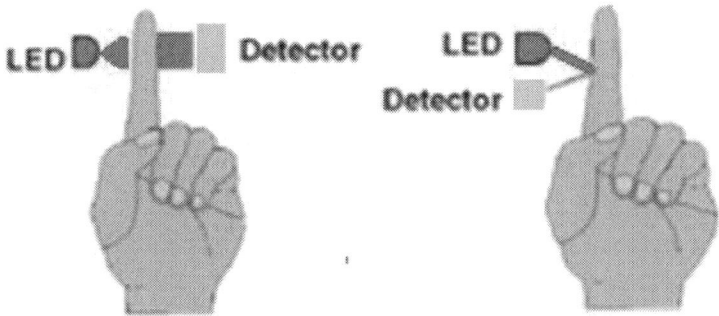

Figure 3. Operation of the pulse oximeter sensor, left: transmission pulse oximetry, right: reflection pulse oximetry

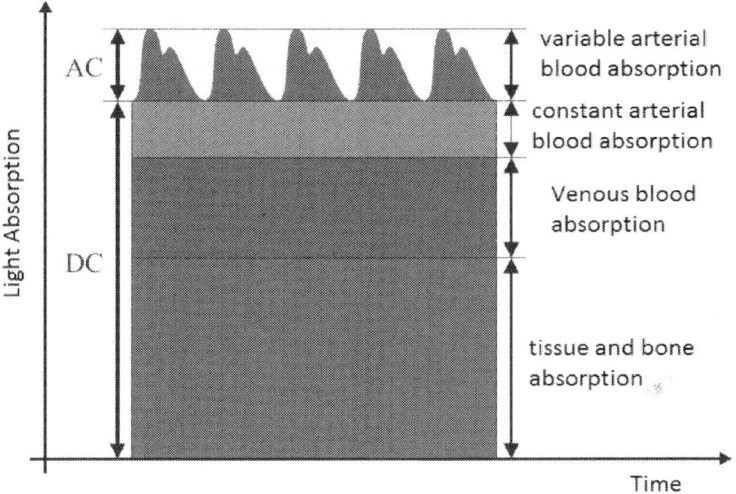

Figure 4. Light absorption by different tissue, at the top we see the plethysmogram generated by the arterial pulsationen

Pulse oximeter works according to two physical principles: first, the presence of a pulse wave generated by changes in blood volume (plethysmography) in the arteries and capillaries (Figure 4) and second, the fact that oxyhemoglobin (O2Hb) and reduced hemoglobin (Hb) have different absorption spectra (spectroscopy). Oxygenated hemoglobin absorbs more infrared light and allows more red light to pass through. Deoxygenated (or reduced) hemoglobin absorbs more red light and allows more infrared light to pass through.

3. ADAPTIVE FILTERING OF PHOTOPLETHYSMOGRAM

We emphasize heir on the use of the adaptive filter by PPG, because of the importance of this signal by detecting further parameters like pulse transit time (PPT); blood pressure monitoring, Pulse rate variability and the application of it for the risk estimation and diagnosis of cardiovascular diseases. Also the non-invasive calculation of concentration, fractional oxygen saturation and further blood components like glucose may require also the PPG signal analyzing. AC component of PPG signal caries important information for diagnosis, but it may be affected by noise, which is sharing the same bandwidth. An important application for the PPG is the calculation of oxygen saturation in emergency and in intensive care, where the oxygen supplement of tissue has to be measured continuously. The problem will be greater for example by detecting the PPG by low perfusion for the monitoring of oxygen saturation, where a low signal to noise ratio is the result. An adaptive filter will be the solution for this problem. Conventional filtering cannot be applied to eliminate those types of artifacts because signal and artifacts have overlapping spectra. For long term monitoring an adaptive filter is essential [Com 2007].

By pulse oximetry, Masimo adaptive filter is well known to the people working in this area. The principle is easy and shortly described here: all detected samples of PPG`s by red and infrared causing oxygen saturation below a certain value (e.g. 80%) are coming from venous blood signals caused by motion artifact and has to be filtered. All signals causing saturation higher than a threshold value (e.g. 90%) are the arterial signal. An intelligent algorithm is designed according to this principle for the robust detection of oxygen saturation. By using one PPG signal we cannot apply this algorithm. We used another algorithm by Filtering and generation of reference noise depending on the detected signal.

Motion artefacts are one of the most important handicaps of photopletysmography and pulse oximetry, as they suppose a big limitation and often become an insurmountable obstacle on the utilization of this technology, since they are quite hard to cancel mainly due to spectral characteristics of both, pulse signals and motion artifacts. In order to improve the quality of Photoplethysmograms and pulse oximetry, some signal processing must be implemented. Our research proposes, as viable solution, an Adaptive Filter in Noise Cancellation configuration, working with a Least Mean Square Algorithm. At the end of the system, we have carried out a reconstruction of the Photoplethysmogram and the signal that we recover has a high enough quality for measuring fractional oxygen saturation of hemoglobin in blood and for further diagnosis purposes.

An Adaptive Noise Cancellation (ANC) System has two inputs. This fact can be seen in the Figure 5presented below, more specifically in the diagram on the top. One is the Input Signal, i.e., the signal corrupted by noise, coming from the sensor output, and the other one is the Noise Reference, coming from the

Synthesizer output. Both, the graphic of the Input Signal and the generated plot of the Noise Reference appear in the Front Panel of the corresponding LabVIEW program. Given that the Least Mean Square Algorithm provides adaptive filtering, the Noise Reference is adjusted to the real noise measured with the sensor and, as a result, the output, Filtered Signal, naturally will be the filtered signal. In the diagram below from the Figure 5 the main blocks of the Least Mean Square Algorithm (LMS) implementation are presented. It is worth mentioning the fact that this algorithm is recursive: the weights of the filter are calculated recursively to minimize the Mean Square Error [Abdallah].

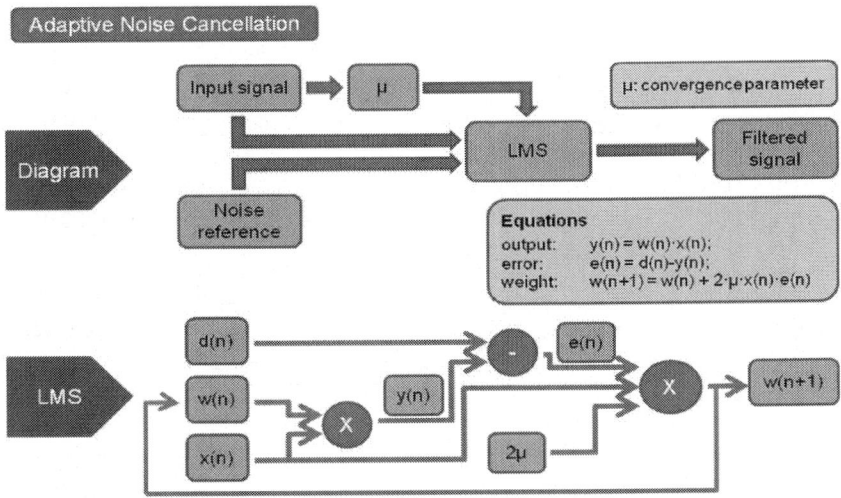

Figure 5. Block diagram of the Adaptive Noise Cancellation (ANC)

4. METHOD AND RESULTS BY ADAPTIVE FILTERING OF PHOTOPLETHYSMOGRAM

Adaptive filters have been used to enable the measurement of photoplethysmogram PPG under conditions, where movement of the body parts where the sensor is applied causes a high noise to the signal. In this adaptive filter a noise reference and a signal reference are used. We use the least mean square (LMS) method to extract the actual signal from the noisy one.

For the first approximation to generate the reference signal a lowpass filter is used. Using the resulting signal from this lowpass, an appropriate reference signal is generated. This reference signal is in turn subtracted from the detected signal to generate a noise signal. The generated noise signal is modified to synthesize the noise reference signal. The synthesized reference noise is adjusted by the adaptive algorithm to the real one contained in the measurement, and then subtracted from the detected noisy signal. The resulting signal is modified to fulfill certain requirements (Figure 6)

The applied method discussed above can be used for the detection of a photopletysmogram signal without the need for further signals of the same type or requiring a further sensor. Figure 7 shows an example of the results obtained using this method. The algorithm was tested for the calculation of oxygen saturation and accurate results are delivered under artificial motion artifacts.

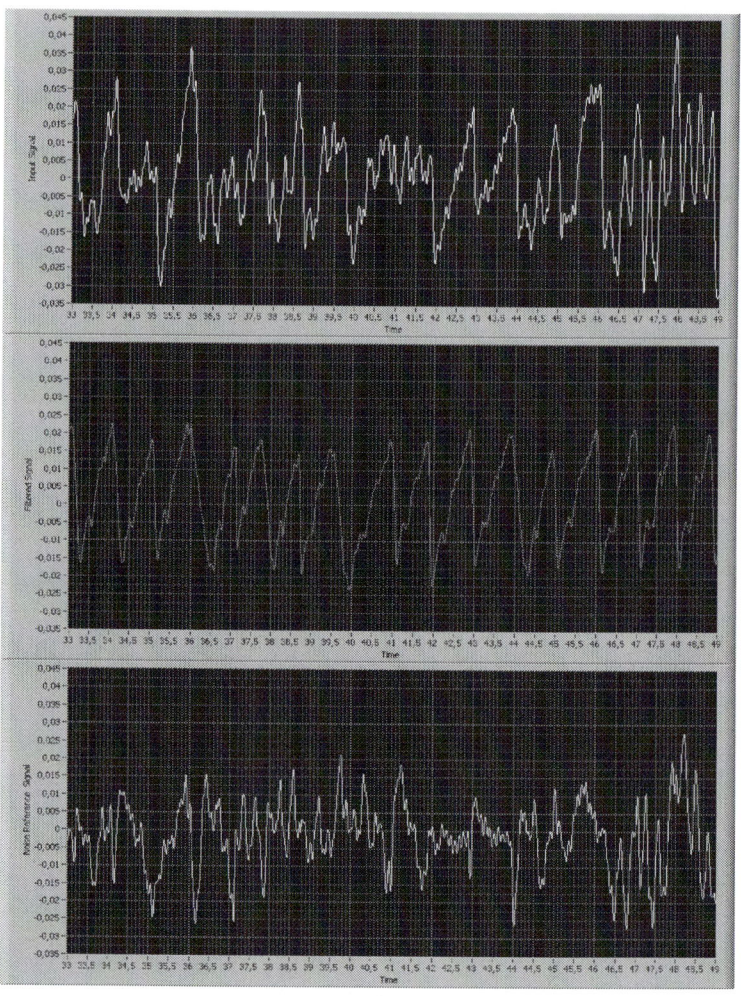

Figure 6. Detected signal (top), generated reference signal (middle) and generated reference noise for PPG filtering

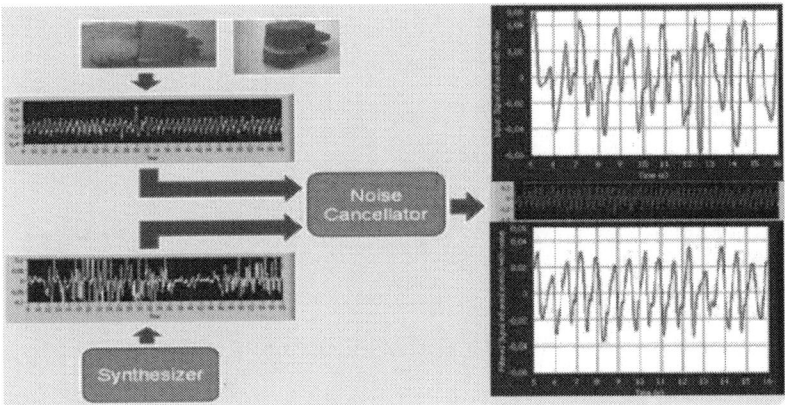

Figure 7. Schematic of the PPG filtering

Each measurement from the applied PHM sensor contains seven signals of LEDs having different wavelengths. Besides, a LED (which is off) acts as zero reference level. Since we need two of them, first we have to separate them. Once these signals are presented separately, we select the two of them that have been measured with the proper wavelengths value for the calculation of oxygen saturation (LED having the wavelength 970 nm, representing infrared light and a LED having the wavelength 660 nm, representing red light). Then they are already adapted for being filtered by our system, which remove the motion artifact from them. Finally, the filtered signals obtained after the program execution can already be used to compute ratios regarding the SpO2, such as the so-called Ω ratio:

$$\Omega = \frac{\ln \dfrac{I(\lambda_1, t_1)}{I(\lambda_1, t_2)}}{\ln \dfrac{I(\lambda_2, t_1)}{I(\lambda_2, t_2)}} \tag{1}$$

Where:

$I(\lambda_1, t_1)$, $I(\lambda_1, t_2)$, $I(\lambda_2, t_1)$ and $I(\lambda_2, t_2)$ are the light intensities measured at the instants t_1, t_2 with the wavelengths λ_1, λ_2 respectively.

As results, examples of each step of the process described here are presented. First of all, examples of the appearance of PHM measurements (and therefore, multiwavelength measurements) are shown, both the whole measurement and a zoom of it (Figure 8).

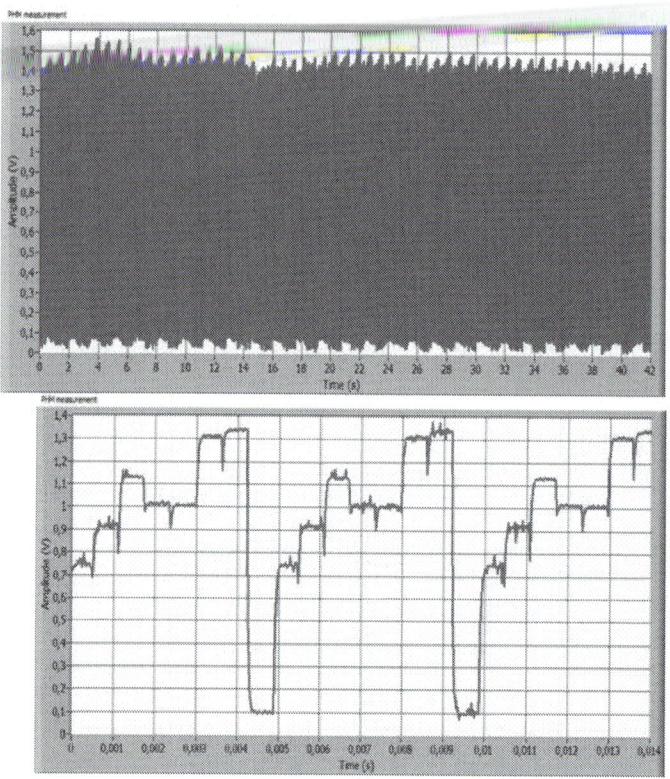

Figure 8. Measurements of photoplethysmogram signals of PHM (right) and a zoom of it (left)

Next, the output given by the recovery of each signal is also presented. To demonstrate the ability of the system presented here to make possible a precise enough computation of the SpO2, we have calculated the value of the above-named Ω ratio for several measurements. In order to make sure that the adaptive filter works well enough to get accurate SpO2 readings, the main goals are: first, to prove that the ratios obtained are included in an acceptable range (bearing in mind that the values of this ratio allow us to estimate the calibration that has to be applied later to the exact calculation of the SpO2). Next, it must be proved that the values for the ratio when the signal is affected by motion artifacts keep quite unchanging compared to those derived from the same signals without motion artifacts [Figure 10]. The pulse amplitudes of the red and infrared signals are detected by the pulse oximeter and measured to produce a certain ratio value, which is intrinsically related to the functional oxygen saturation of (SpO2).

The signals shown in Figure 8 are measured by a Pulshemometer (PHM) sensor for the aim of calculation of concentration and fractional oxygen saturation SaO2, which based on the Principle of plethysmography (here volume change of arterial blood due to pulsation generated from the heart) and optical spectroscopy. Also by our Project for the non-invasive monitoring of glucose concentration in blood an adaptive filter for this aim is essential. For in vivo measurement of blood components, the adaptive filter is necessary to get rid of the noise and disturbances to the signal without any distortion of the detected useful signal that may cause erroneous additive signals or that may reduce the information contents in the detected signal. The Pulshemometer PHM sensor with a compact hardware circuit for driving the LED's and programmable digital potentiometer for adaptive programmable gain amplification is shown in Figure 9.

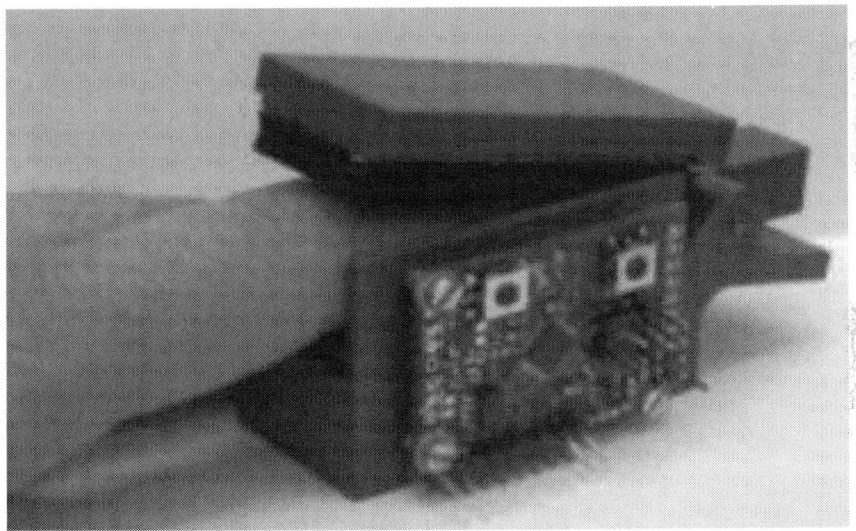

Figure 9. Pulshemometer PHM sensor for hemoglobin concentration and fractional and fractional

Seven separated filtered PHM signals for and fractional oxygen saturation measurement von PHM are shown in Figure 11 after normalization. For this sensor an adaptive filter is essential for reliable and high accuracy results.

Wavelet transformation in combination with fuzzy and neuronal Networks (in some cases cascaded) adaptive filtering is applied by different research groups. An energy ratio-based method and a wavelet-based cascaded adaptive filter (CAF) can be applied for detecting and removing baseline drift from pulse waveforms. This CAF outperforms traditional filters both in removing baseline drift and in preserving the diagnostic information of pulse waveforms [Lisheng]. Daubechies wavelet adaptive filter based on Adaptive Linear Neuron networks is used to extract the signal of the pulse wave. Wavelet transform is a powerful tool to disclose transient information in signals. The wavelet used is adaptive because the parameters are variable, and the neural network based adaptive matched filtering has the capability to "learn" and to become time-varying. So this filter estimates the deterministic signal and removes the uncorrelated noises with the deterministic signal. This filter is found to be very effective in detection of symptoms from pulsatile part of the entire optical signal [Xiaoxia]. Fuzzy logic and Neuro-fuzzy can be used by adaptive filtering.

The method that has to be applied depends on the sensor applications and the case under consideration, because intensive computation time, a high speed processor and a large saving space may be needed, which may cause a delay time that disables an online monitoring. In applications by multi-monitoring it will be possible to use other detected signals for the purpose of filtering of a certain signal as will be discussed on the following section.

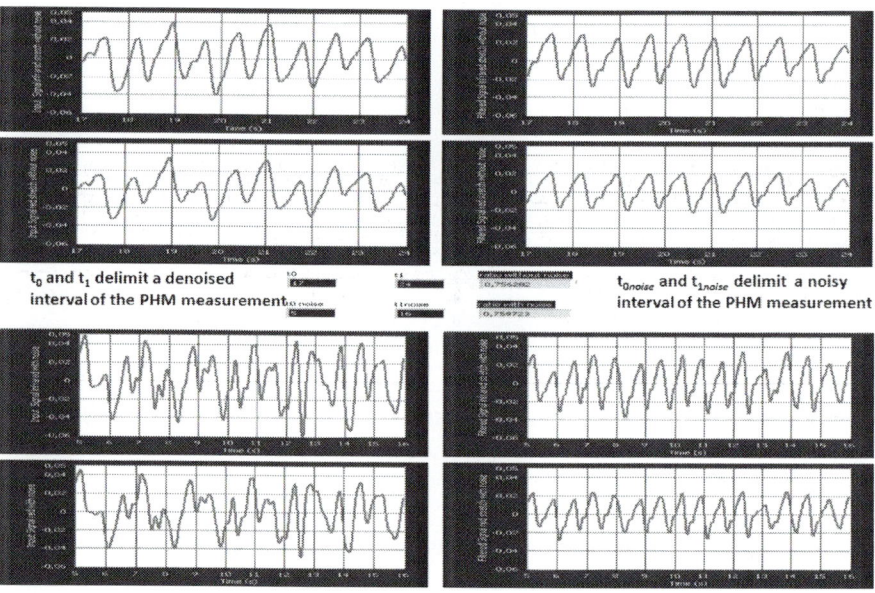

t_0 and t_1 delimit a denoised interval of the PHM measurement

t_{0noise} and t_{1noise} delimit a noisy interval of the PHM measurement

Figure 10. LabVIEW printing for filtering a PHM measurement and computing Ω

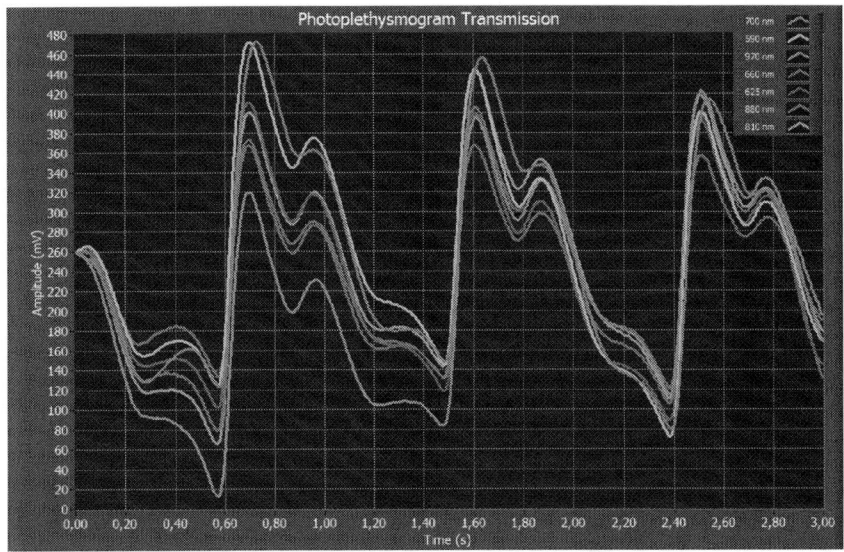

Figure 11. LabVIEW printing for 7 filtered PHM signals for and fractional oxygen saturation measurement

5. APPLICATION OF FURTHER SIGNALS FOR ADAPTIVE FILTER OF PHOTPLETHYSMOGRAM

In one method is to use a simple acceleration sensor of a piezoelectric element [Lit. Han2009]. In a current work we use also a method based on adaptive filtering by taking the advantage of piezoelectric sensor signal to get information about the desired signal or motion to estimate the motion artifact noise or to generate the reference signal in order to get the filtered one. Hence the piezoelectric signal describes the velocity in the blood volume change; numerical integration of the piezoelectric element signal may be helpful by synthesizing of reference signal or the noise that has to be subtracted from the noisy detected signal by generating artificial motion artifacts after signal processing. Figure 12 shows the simultaneously detected optical signal detected on fingertip and a piezoelectric signal detected on the wrist with artificial motion artifact in the time period between 11 and 15 second. The piezoelectric signal is still clean and can be used for the generating of the optical signal.

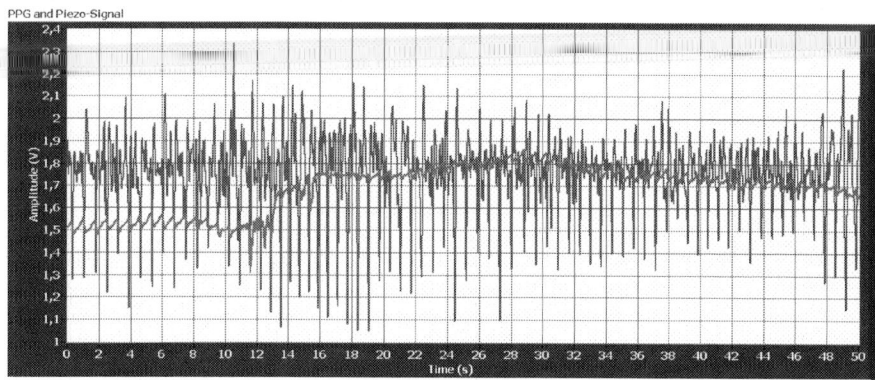

Figure 12. PPG and piezosignal by motion artifact

Figure 13 shows the simultaneously detected optical signal detected on fingertip and a piezoelectrical signal in the case of extrasystole also detected on the wrist without motion artifact.

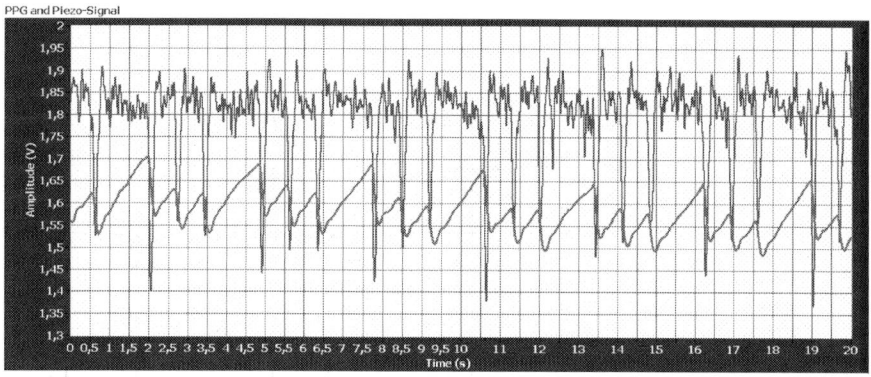

Figure 13. PPG and piezosignal by extrasystole

Figure 14 shows the simultaneously detected optical signal detected on fingertip and the piezoelectrical signal detected on the wrist with artificial motion artifact. The piezosignal is still good enough for the generating of the PPG signal.

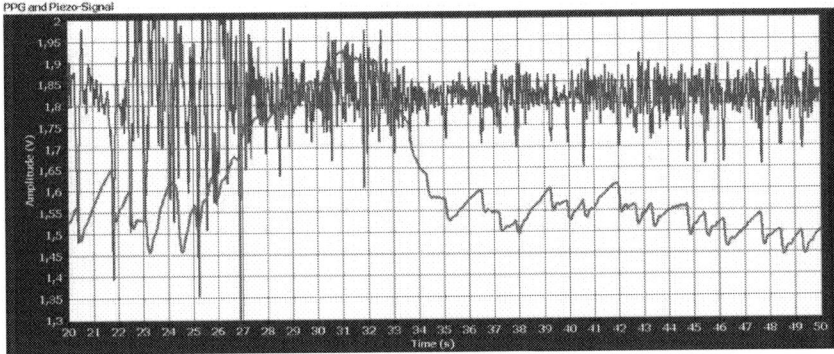

Figure 14. PPG and piezosignal by extrasystole under motion artifact

The piezoelectric sensor may be also used for the measurement of other physiological parameters when PPG or ECG is also available. Pulse transit time PTT, which has shown its potential in cardiovascular studies, can be calculated from ECG and piezoelectric signal detected on the radial artery instead of PPG signal by low perfusion [Abdiel]. Under normal conditions the PTT can be calculated using a simple and compact piezoelectric sensor and a PPG one or by using only two simple, compact, low power piezoelectric elements. Piezoelectric elements have also different applications in biomedical engineering as sensors and actuators [Fannin].

Also a light source having the wave length around 590 nm can be used to generate signals that are larger than that of other wavelengths due to the high light absorption of hemoglobin at this wavelength.

Motion artifacts effects on the signal quality of photoplethysmographic signals are also discussed by different research groups. Using 3-D acceleration sensors are applied and showed very good results for adaptive filtering. The focus by [Volmer] lies on a performance estimation for a reconstruction method based on adaptive filtering with help of acceleration signals acquired at the fingertip. The acceleration in the direction of finger is an optimal setup for a continuous long-term application with low cost and low calculation complexity.

6. ADAPTIVE FILTERING BY FURTHER BIOSIGNALS

The previous discussed method for the removal of the motion artifact from a PPG signal can be applied by other biosignals like ECG or EEG. We discussed in this chapter briefly the adaptive filtering by other biosignals described by different authors as examples to emphasize on the importance of adaptive filtering. Figure 15 shows a 3-chanel ECG, subjected at the end portion to motion artifact. Low signal to noise ratio can make the task of an adaptive filter more difficult.

An automated system for efficient detection of brain tumors in EEG signals using artificial neural networks (ANNs) is described by Murugesan and Sukanesh [Murugesan]. Generally, the EEG signals are bound to contain an assortment of artifacts from both subject and equipment interferences along with essential information regarding abnormalities and brain activity. Adaptive filtering has to be applied to remove the artifacts present in the EEG signal.

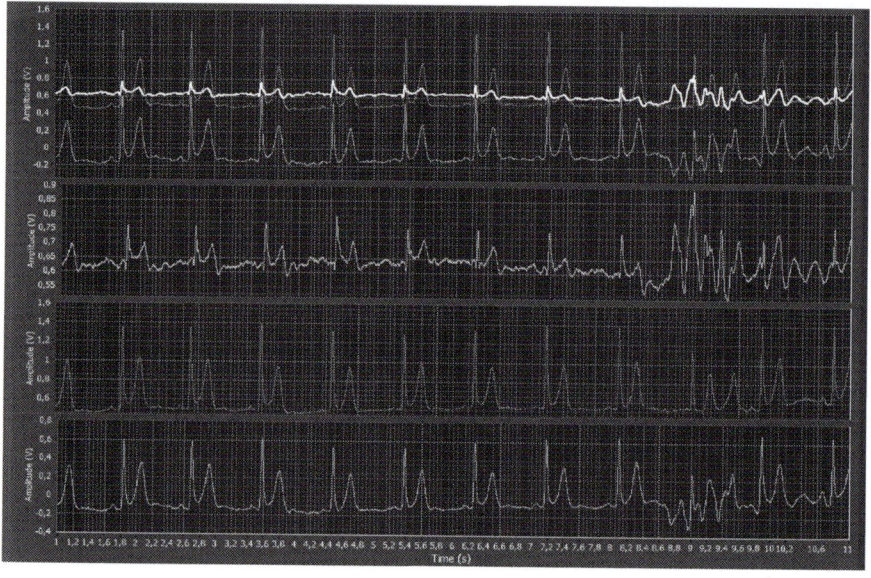

Figure 15. ECG with motion artifact by the last three seconds

Artifacts in EEG records are caused by various factors, like line interference, EOG and ECG. These noise sources increase the difficulty in analyzing the EEG and to obtaining clinical information. For this reason, it is necessary to design specific filters to decrease such artifacts in EEG. A cascade of three adaptive filters based on a least mean squares (LMS) algorithm will be helpful. The first one eliminates line interference, the second adaptive filter removes the ECG artifacts and the last one cancels EOG spikes [Correa].

The MEG signal can also be used to control rehabilitation systems like prostheses and artificial neuromuscular electrical stimulation toward restoring movement to spinal cord injured patients. These mobile systems are usually used in different environments and thus are being exposed to different noise levels with characteristics not completely known. Different techniques for noise reduction have been compared for that aim: wavelet transform (WT), adaptive digital filters, and non-adaptive digital filters [Ortolan].

By non-invasive hemodynamic measurements like stroke volume and cardiac output, impedance cardiography (ICG) can be applied for continuously measurements. The impedance cardiography is designed for assessment and management of congestive heart failure, hypertension, and pacemaker patients. It is also appropriate for select ICU patients. The ICG waveform is generated by thoracic electrical bioimpedance (TEB) technology, which measures the level of change in impedance in the thoracic fluid [Philips]. Adaptive Filtering is in this case essential for reliable measurement at different conditions. Impedance plethysmography is a well known and extensively used noninvasive method for physical parameters monitoring. A major problem with this technique is its sensitivity to body movement. More than one frequency impedance plethysmograph can be used to measure different body parts movements. Adaptive filtering is heir very essential to detect and reduce motion artifacts [Rosell]. Wavelet based cancellation of artifacts in impedance cardiography is discussed by Pandey [Pandey].

Phonocardiography, Ballistocardiography may be also important for the extracting more information of the cardiovascular system as well as the low frequency signals caused by sympathic and parasympathic autonomic nervous system for the diagnostic of cardiovascular diseases and its autoregulations [Xinsheng].

Contactless measurement for physiological parameter like ECG, EEG [Oehler] and respiration rate is research area by different groups, where adaptive filtering is of great importance. Cardiac pulse based on the information contained in the thermal signal emitted from major superficial vessels is discussed by Pavlidis et. al. [Garbey]. Non-contact monitoring of breathing function using infrared imaging is also discussed by R. Murthy and I. Pavlidis.

At the end of this chapter we want only to indicate that by image processing in biomedical engineering the adaptive filtering is very essential and discussed in many literatures [Sudha].

In summary for both, long-term and short-term monitoring (for example in case of an emergency or for prevention) of biosignals, the use of an adaptive filter is essential for reliable results.

7. CONCLUSIONS

The motion artifact caused by subject motion as well as by physical motion of body parts has a varying frequency which may lies in the same range of the signal frequency. It is difficult to filter noise from these signals using traditional filters, and errors resulting from filtering can distort them and physicians may misled by these noisy signals and this may make the diagnosis not possible or an erroneous diagnosis is the result. By intensive care and emergency the continuous supplement of tissue with the oxygen is crucial (especially brain tissue, where after few minutes oxygen deficiency causes irreversible damage of tissue). The calculation of oxygen saturation by pulse oximetry is based on photoplethysmography and transmission microscopy. Hence the photplethysmogram plays a great rule by the calculation of oxygen saturation and heart rate as well as many further diagnostic parameters like stiffness, blood pressure and more cardiovascular pathologies can be calculated using other parameters like impedance cardiography and ECG. For a reliable detection of these parameters, an adaptive filtering is essential. By oxygen saturation detection, since many years an adaptive filter from the company Masimo that accepts all measurement results for the calculation of oxygen saturation above a threshold value and reject all measurements yielding results below another value, is robust and in clinical use. By our method using a tiefpass with a low cut off frequency a robust results were obtained. The potential of this method is that only one signal is required, which enables its application by compact sensors for different applications. Also a high velocity in the ascending portion of a signal followed by exponential or polynomial decay may consider as a PPG signal. In multi-monitoring or using multisensory for diagnosis, another signals can be used by adaptive filtering. As an example ECG or piezoelectric signals can be applied for the robust detection of PPG. For its simplicity and cost effectiveness as well as its quickly application without great disturbance of the patient we use piezoelectric signals. Acceleration sensors find their application by the adaptive filtering for PPG or ECG and further physiological parameters. By multisensors some signals can be used to generate the reference signal that has to be extracted from the measured signal or the reference noise that has to be removed from the measured data. Kalman Filter, Neuronal Fuzzy with optimization methods like swarm algorithms and Wavelet Transformation may be in some cases a good choice for adaptive filtering of biosignals. The method which has to be applied depends on the case under consideration and the availability of other sensors. For emergency, intensive care, home care and long term monitoring and over all, where non-invasive measurement are applied, the use of adaptive filter is of a great importance and in many cases is compulsory to get the required results. It will also radically reduce the disturbances (alarm) for patient and medical care stuff, reduce costs and enhance the medical systems.

REFERENCES

1. O. Abdallah, Tarazona. A. Piera, Roca. T. Martínez, H. Boutahir, Alam. K. Abo, A. Bolz, 2006 Photoplethysmogram Signal Conditioning by Monitoring of Oxygen Saturation and Diagnostic of Cardiovascular Diseases, 4th European Congress for Medical and Biological Engineering, 978-3-54089-207-6 303306), Antwerp, September 2008,

2. Foo Jong Yong Abdiel, Chu Sing Lim , 2006 Pulse Transit Time based on Piezoelectric Technique at the redial Artery. Journal of clinical monitoring and computing, (May 2006) 20 Nr. 3, 185192

3. Jan-Michael Abicht, 2003 Computerunterstuetzte Analyse photoplethysmographischer Signale, Dissertation zum Erwerb des Doktorgrades der Medizin an der Medizinischen Fakultaet der Ludwig-Maximilians Universitaet zu München, October 2003, available from: http://edoc.ub.uni-muenchen.de/1793/1/Abicht_Jan_Michael.pdf

4. AllenJohn 2007 Photoplethysmography and its application in clinical physiological Measurement, Physiological Measurement 28,, (February 2007), R1R39

5. G. Comtois, Y. Mendelson, P. Ramuka, 2007 A Comparative Evaluation of Adaptive Noise Cancellation Algorithms for Minimizing Motion Artifacts in a Forehead-Mounted Wearable Pulse Oximeter, Conf proceeding IEEE Eng Medicine Biology Soc EMBS, 978-1-42440-787-3 15281531 , Lyon, August 2007

6. Christopher. A. Fannin, 1997 Design of an Analog Adaptive Piezoelectric Sensoriactuator, Thesis submitted to the Faculty of the Virginia Polytechnic Institute and State University in partial fulfillment of the requirements for the degree of MASTER OF SCIENCE, 1997 From :http://schoöar.lib.vt.edu/thesis/etd-8897-171952/unrestrictd/Cfannin.pdf

7. Garbey Marc Sun Nanfei, Merla Arcangelo, Pavlidis Ioannis, . 2007 Contact-Free Measurement of Cardiac Pulse Based on the Analysis of Thermal Imagery, IEEE Transactions On Biomedical Engineering, 54 8 August 2007, 14181426

8. D. K. Han, J. H. Hong, J. Y. Shin, T. S. Lee, 2009 Accelerometer based motion noise analysis of ECG signal, World Congress on Medical Physics and Biomedical Engineering,, IFMBE Proceedings, 25 5, 198201 , Munich, Germany, September 2009

9. Ju-Won Lee, Gun-Ki Lee, 2005 Design of an Adaptive Filter with a Dynamic Structure for ECG Signal Processing, International Journal of Control, Automation, and Systems, 3 1 (March 2005), 137142 ,

10. M. Lichong, Z. David, W. Kuanquan, 2005 Wavelet-Based Cascaded Adaptive Filter for Removing Baseline Drift in Pulse Waveforms, IEEE Transactions on Biomedical Engineering, 52 11 (November 2005), 19731975 , 0018-9294

11. M. Murugesan, R. Sukanesh, 2009 Towards Detection of Brain Tumor in Electroencephalogram Signals Using Support Vector Machines, International Journal of Computer Theory and Engineering, 1 5 (December, 2009), 17938201

12. M. Murugesan, R. Sukanesh, 2009 Automated Detection of Brain Tumor in EEG Signals Using Artificial Neural Networks, Int. Conf. on Advances in Computing, Control, and Telecommunication Technologies, 284288 , Trivandrum, India, December 2009

13. OehlerMartin Johannes 2009 Kapazitive Elektroden zur Messung bioelektrischer Signale, Technischen Universitaet Carolo-Wilhelmina zu Braunschweig, Dissertation 2009 Available from: http://rzbl04.biblio.etc.tu-bs.de:8080/docportal/receiv/DocPortal_document_00031116

14. 16. R. L. Ortolan, Pereira. R. R. Mori, C. M. Cabral, J. C. Pereira, A. J. Cliquet, 2003 Evaluation of adaptive/nonadaptive filtering and wavelet transform techniques for noise reduction in EMG mobile acquisition equipment, Neural Systems and Rehabilitation Engineering, IEEE Transactions 11 1 (March 2003) 6069 , 1534-4320

15. Vinod. K. Pandey, Prem. C. Pandey, 2007 Wavelet based cancellation of respiratory artifacts in impedance cardiography, IEEE Intl. Conf. on Digital Signal Processing, Cardiff, Wales, UK, July 2007

16. Philips Healthcare: ICG Impedance Cardiography, Non-invasive hemodynamic measure-ments,http://www.healthcare.philips.com/main/products/patient_monitoring/products/icg/

17. D. V. Prasad, R. Swarnalatha, 2009 A New Method of Extraction of FECG from Abdominal Signal, Int. Conf. On Biomedical Engineering, IFMBE Proceedings, 23 98100 , Singapore, December 2008

18. Rasheed Tahir, Ho In Myung, Lee Young-Koo, Lee Sungyoung, Lee Soo Yeol, Kim Tae-Seong, 2006 Constrained ICA Based Ballistocardiogram and Electro-Oculogram Artifacts Removal from Visual Evoked Potential, EEG Signals Measured Inside MRI, Lecture Notes in Computer Science, 4232 2006, 10881097 ,

19. Rik Vullings, Chris Peters, Massimo Mischi, Rob Sluijter, Guid Oei, Jan Bergmans, 2007 Artifact reduction in maternal abdominal ECG recordings for fetal ECG estimation, Proceedings of the 29th Annual International Conference of the IEEE EMBS, Lyon, France, August 2007

20. Javier. Rosell, Kevin. P. Cohen, John. G. Webster, 1995 Reduction of motion artifacts using a two-frequency impedance plethysmograph and adaptive filtering, IEEE Transactions On Biomedical, Engineering, 42 10 (October 1995), 10441148 , 0018-9294

21. S. Sudha, G. R. Suresh, R. Sukanesh, 2009 Speckle Noise Reduction in Ultrasound Images by Wavelet Thresholding based on Weighted Variance, International Journal of Computer Theory and Engineering, 1 1 April 2009, 17938201 ,

22. T. Rasheed, L. Young-Koo, L. Y. Soo, T. S. Kim, 2009 Attenuation of artifacts in EEG signals measured inside an MRI scanner using constrained independent component analysis, Physiol. Meas., 30 4 April 2009, 387404

23. Achim Volmer, Reinhold Orglmeister, Sebastian Feese, 2010 Motion Artifact Compensation for Photoplethysmographic Signals by Help of Adaptive Noise Cancelation Motion, Automatisierungstechnik, 58 5 May 2010, 269276

24. Xiaoxia Li, Gang Li, Ling Lin, Yuliang Liu, Yan Wang, Yunfeng Zhang, 2004 Application of a Wavelet Adaptive Filter Based on Neural Network to Minimize Distortion of the Pulsatile Spectrum, Advances in Neural Networks Lecture Notes in Computer Science, 3174 2004, 279301 , ISNN 2004

25. Yu Xinsheng , Don Dent, Colin Osborn, . 1996 Classification of Ballistocardiography using Wavelet Transform and Neural Networks, Annual International Conference of the IEEE Engineering in Medicine and Biology Society, 3 October 1996, 937938 , 0-78033-811-1

26. Y. S. Yan, C. C. Poon, Y. T. Zhang, 2005 Reduction of motion artifact in pulse oximetry by smoothed pseudo Wigner-Ville distribution, Journal of NeuroEngineering and Rehabilitation, 2 : 3 EOF , March 2005,

Chapter 8

Adaptive Filters for Processing Water Level Data

Natasa Reljin[1], Dragoljub Pokrajac[1] and Michael Reiter[1]

[1] Delaware State University, USA
[2] Bethune-Cookman University, USA

1. INTRODUCTION

Salt marshes are composed of various habitats contributing to high levels of habitat diversity and increased productivity (Kennish, 2002; Zharikov et al., 2005), making them among the most productive ecosystems on the Earth. The salt marsh consists of a halophytic vegetation community growing near saline waters (Mitsch & Gosselink, 2000) characterized by grasses, herbs, and low shrubs (Adam, 2002). Salt marshes exist between the upper limit of the high tide and the lower limit of the mean high water tide (Adam, 2002). They represent an important factor in the support of surrounding food chains, and due to the high level of productivity their economic and aesthetic value is increasing (Delaware Department of Natural Resources and Environmental Control, 2002; Zharikov et al. 2005). The survival and reproduction of many species of commercial fish and shellfish is dependent upon salt marshes (Zharikov & Skilleter, 2004). In addition, salt marshes provide critical habitat and food supply to crustaceans (Zharikov et al., 2005) and shorebirds (Potter et al., 1991). They are often considered as a primary indicator of the ecosystem health (Zhang et al., 1997). Because of their ability to transfer and store nutrients, salt marshes are an important factor in the maintenance and improvement of water quality (Delaware Department of Natural Resources and Environmental Control, 2002; Zhang et al., 1997). In addition, they provide significant economic value as a cost-effective means of flood and erosion control (Delaware Department of Natural Resources and Environmental

Control, 2002; Morris et al., 2004). This economic value makes coastal systems the site of elevated human activity (Kennish, 2002).

Determining the effects of sea level rise on tidal marsh systems is currently a very popular research area (Temmerman et al., 2004). While average sea level has increased 10-25 cm in the past century (Kennish, 2002), the Atlantic coast has experienced a sea level rise of 30 cm (Hull & Titus, 1986). Local relative sea level has risen an average rate of 0.12 cm yr^{-1} in the past 2000 years, but at Breakwater Harbor in Lewes, DE sea level is rising at the average rate of 0.33 cm yr^{-1}, nearly three times that rate (Kraft et al., 1992). According to the National Academy of Sciences and the Environmental Protection Agency, sea level rise within the next century could increase 60 cm to 150 cm (Hull & Titus, 1986).

The changes in sea level rise are particularly affecting tidal marshes, since they are located between the sea and the terrestrial edge (Adam, 2002; Temmerman et al., 2004). The prediction is that sea level rise will have the most negative effect on marshes in the areas where the landward migration of the marsh is restricted by dams and levees (Rooth & Stevenson, 2000). If sea level rises the almost certain prediction of 0.5 m by 2100 and marsh migration is prevented, then more than 10,360 km^2 of wetlands will be lost (Kraft et al., 1992). If the sea level rises 1 m then 16,682 km^2 of coastal marsh will be lost, which is approximately 65% of all extant coastal marshes and swamps in the United States (Kraft et al., 1992).

Due to an imminent potential threat which can jeopardize the Mid-Atlantic salt marshes, it is very important to examine the effect of sea level rise on these marshes. The marshes of the St. Jones River near Dover, DE, can be considered to be typical Mid-Atlantic marshes. These marshes are located in developing watersheds characterized by dams, ponds, agricultural lands, and increasing urbanization, providing an ideal location for studying the impacts of sea level rise on salt marsh extent and location. In order to determine the effect of sea level rise on the salt marshes of the St. Jones River, the change in salt marsh composition was quantified. Unfortunately, as for most marsh locations along the Atlantic seaboard, the data on sea level rise for this area was not available for comparison with marsh condition. However, a wide data set for this area is available through a water quality monitoring program, and if it could be properly processed and analyzed it could result in sea level rise data for the location of the interest.

In this chapter, we describe the application of signal processing on the water level data from the St. Jones River watershed. The emphasis is on adaptive filtering in order to remove the influence of upstream water level on the downstream levels.

2. DATA

The St. Jones River, in central Delaware, is 22.3 km long (Pokrajac et al., 2007a). It has an average mean high water depth (MHW) of 4 m in the main stem, and an average width of 15 feet. The site's watershed area is 19,778 ha, and the tidal reaches are influenced by fresh water runoff from the urbanized area upstream. An aerial photo of the St. Jones River is shown in Fig. 1.

Figure 1. Aerial photo of St. Jones River.

The data used in this research were obtained from the Delaware National Estuarine Research Reserve (DNERR), which collected the data as part of the System Wide Monitoring Program (SWMP) under an award from the Estuarine Reserves Division, Office of Ocean and Coastal Resource Management, National Ocean Service, and the National Oceanic and Atmospheric Administration (Pokrajac et al. 2007a, 2007b). Through SWMP, researchers collect long term water quality data from coastal locations along Delaware Bay and elsewhere in order to track trends in water quality.

The original dataset contained 57,127 measurements, taken approximately every thirty minutes using YSI 6600 Data Probes (Fig. 2) (Pokrajac et al., 2007a, 2007b). The measurements were taken from January 31, 2002 through October 31, 2005. In order to determine if sea level rise is influencing the St. Jones River, the water level data were collected from two SWMP locations: Division Street and Scotton Landing (Pokrajac et al., 2007b). Probes were left in the field for two weeks at a time, collecting measurements of water level, temperature $(^\circ C)$, specific conductivity (mS cm⁻¹), salinity (ppt), depth (m), turbidity (NTU), pH (pH units), dissolved oxygen percent saturation (%), and dissolved oxygen concentration (mg L⁻¹). We used only the water level (depth) data for this study, which were collected using a non-vented sensor with a range from 0 to 9.1 m, an accuracy of 0.18 m, and a resolution of 0.001 m. Due to the fact that the probes are not vented, changes in atmospheric pressure appear as changes in depth, which results in an error of approximately 1.03 cm for every millibar change in atmospheric pressure (Mensinger, 2005). However, the exceptionally large dataset (57,127 data points) overwhelms this data error.

© Fondriest Environmental, Inc.

Figure 2. YSI 6600 Data Probe.

The downstream location, Scotton Landing, is located at coordinates latitude 39 degrees 05' 05.9160" N, longitude 75 degrees 27' 38.1049" W (Fig. 3). It has been monitored by SWMP since July 1995. The average MHW depth is 3.2 m, and the river is 12 m wide (Mensinger, 2005). This location possesses a clayey silt sediment with no bottom vegetation, and has a salinity range from 1 to 30 ppt. The tidal range is from 1.26 m (spring mean) to 1.13 m (neap mean). The data collected at the Scotton Landing site are referred as *downstream data* (see Fig. 4).

The water level data from the Scotton Landing site alone were not sufficient. In addition to tidal forces, this site is influenced by upstream freshwater runoff, so changes in depth could not be isolated to sea level change. However, the data from a non-tidal upstream sampling site could be used for removing the upstream influence at Scotton Landing. Therefore, the data from an upstream location, Division Street, was included in the analysis. Its coordinates are latitude 39 degrees 09' 49.4" N, longitude 75 degrees 31' 8.7" W (see Fig. 3.). The Division Street sampling site is located in the mid portion of the St. Jones River, upstream from the Scotton Landing site. At this location, the river's average depth is 3 m and width is 9 m. The site possesses a clayey silt sediment with no bottom vegetation, and has a salinity in the range from 0 to 28 ppt. The tidal range at this location varies from 0.855 m (spring mean) to 0.671 m (neap mean). The data were monitored from January 2002 (Mensinger, 2005). The data collected at the Division Street site are referred to as *upstream data* (see Fig. 4).

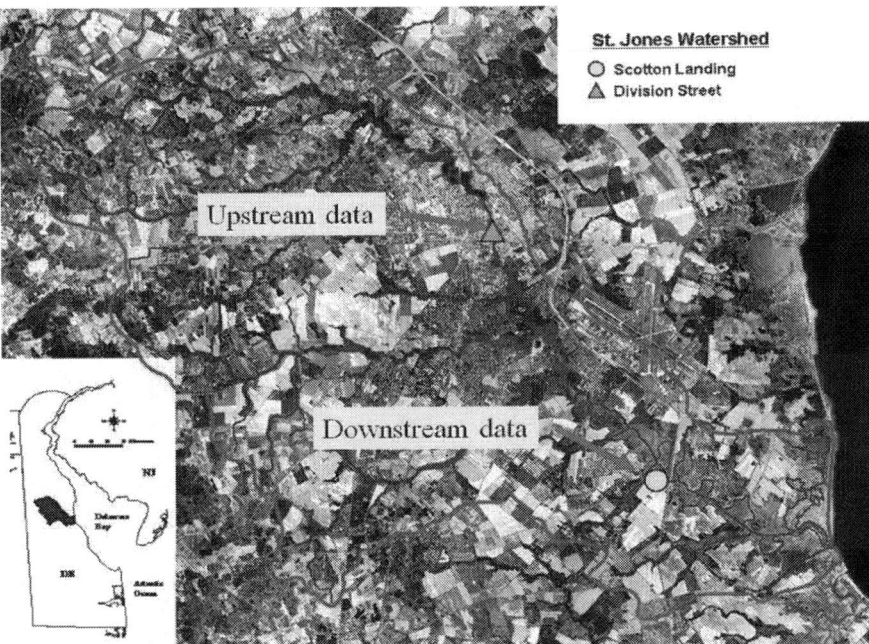

Figure 3. Sampling locations for St. Jones data: Division Street (upstream data); Scotton Landing (downstream data).

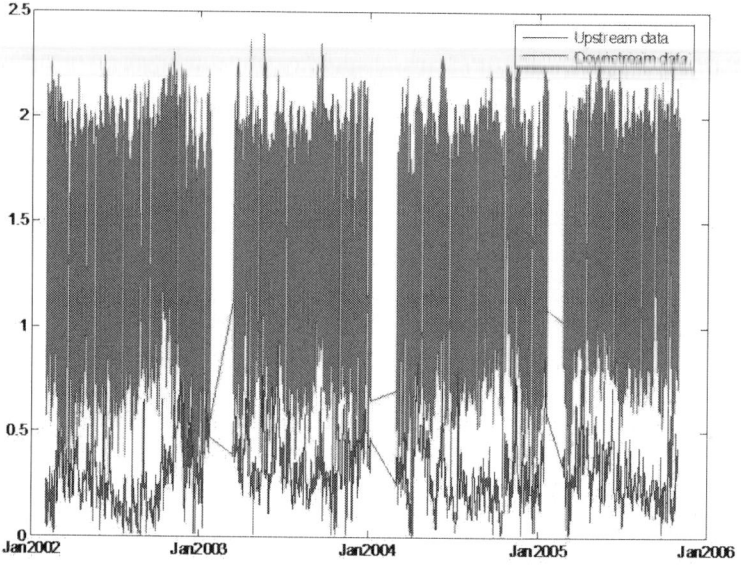

Figure 4. Original dataset (upstream and downstream data).

3. DATA PRE-PROCESSING

The data were sampled every Ts = 30 minutes, and the dataset consisted of "chunks" of continuous measurements. Some of the measurements were missing due to maintenance or malfunction of the equipment, probe replacement, etc. The length of the intervals with missing measurements varied between 1 h (1 missing measurement) and 1517.5 h (3036 missing measurements), but the majority of the intervals were shorter than 10 h.

The discrete Fourier spectra (Proakis & Manolakis, 2006) of all the chunks contained three prominent peaks, which is shown in Fig. 5 using chunk 99 from the downstream data. The first peak corresponds to lunar semi-diurnal tides with a period of approximately 12.4 h, and the diurnal tides with a period of approximately 24.8 h. In addition, there is a peak that corresponds to solar tides, which have a period of approximately 12 h. These periodicities are also shown in Fig. 6.

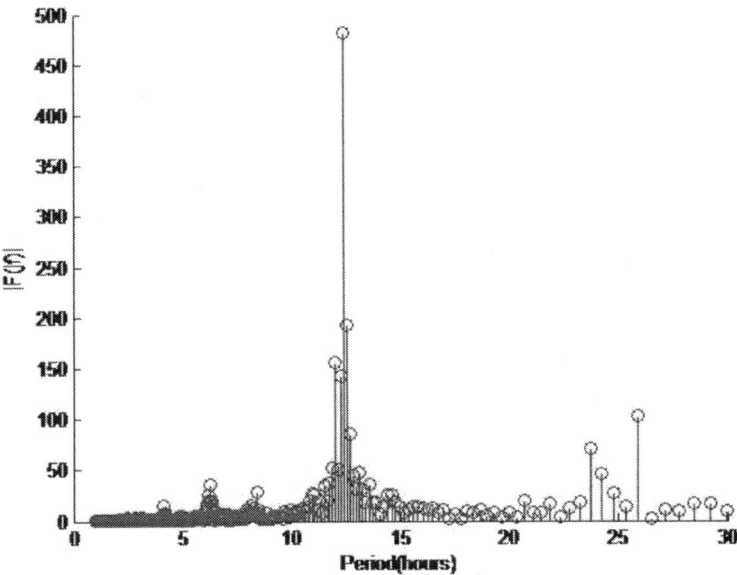

Figure 5. Spectrum of collected data before filtering (chunk 99, downstream data).

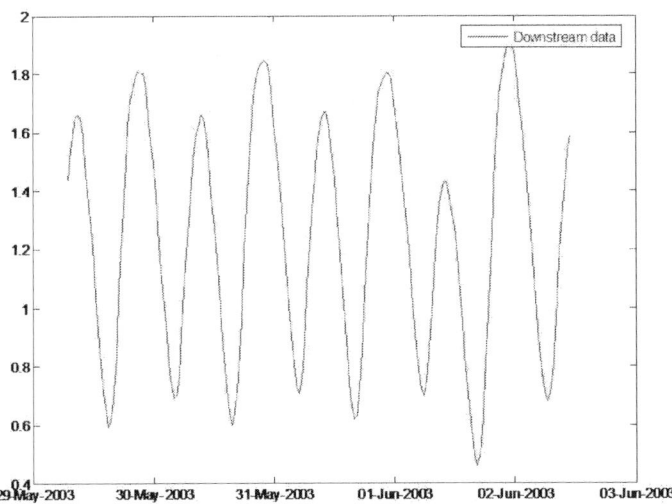

Figure 6. The periodicities of the downstream data.

The dataset had several problems that had to be rectified before further processing. One data sample (Sep 28, 2004, 09:00:00) had an incorrect time, which was located sometime between Sep 27, 2004, 23:30:00 and Sep 28, 2004, 00:30:00, and was corrected. Four data samples (Jul 24, 2003, 07:30:00; Jun 10, 2005, 09:00:00; Aug 11, 2005, 15:00:00; Aug 11, 2005, 15:30:00) had missing values. In addition, the number of intervals with no measurements (total of 99 "gaps" in experiment) represented a problem for signal processing (for example, for filtering). Fig. 7 shows the number of chunks as a function of the duration of the missing measurements. Due to the properties of the used data and the shortest period of 12 h, we decided to interpolate intervals shorter than 12 h. Also, we interpolated all the above mentioned samples with missing data values. The treatment of the missing values is shown in Fig. 8.

In order to interpolate data for each interval of missing measurements, first we approximated the existing data within 20 samples from the interval. We used a least squares approximation followed the combination of the 4th order polynomial and trigonometric functions:

$$x\left(t\right) = \sum_{j=0}^{4} a_j t^j + \sum_{j=1}^{3} A_j \sin\left(\frac{2\pi t}{T_j} + \theta_j\right) \tag{1}$$

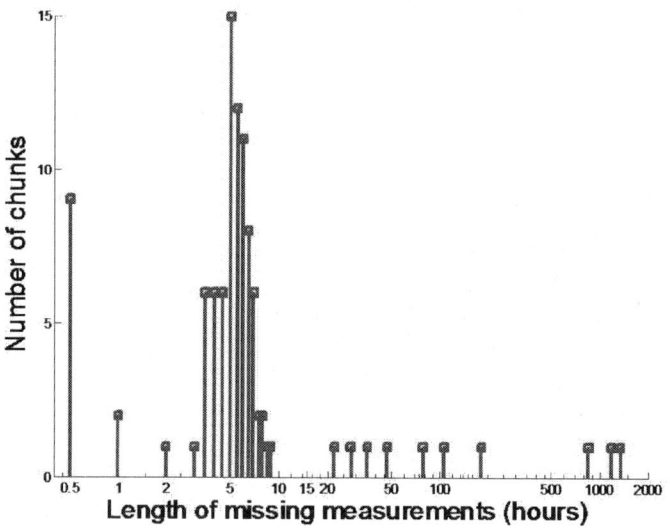

Figure 7. The number of chunks as function of the duration of missing measurements.

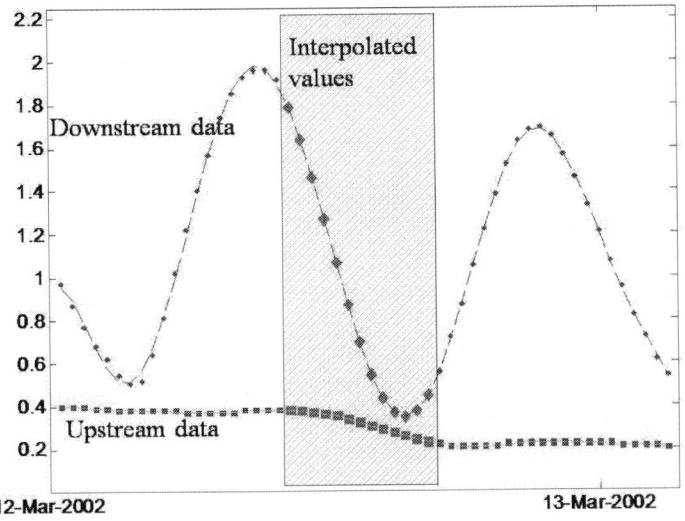

Figure 8. The treatment of the missing values.

where T_1 = 12.4 h, T_2 = 24.8 h and T_3 = 12 h. Then, we interpolated missing values using the computed approximation functions. The interpolation was performed on 866 samples, which represented less than 2% of the original number of samples. One example of the interpolated intervals is depicted in Fig. 9.

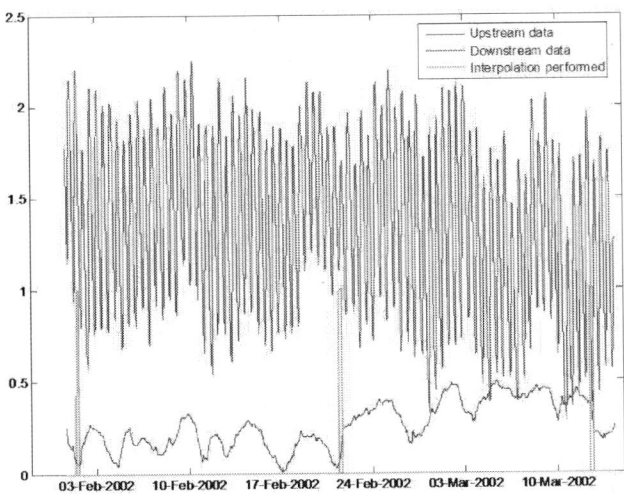

Figure 9. An example of interpolated intervals.

The interpolation resulted in the merging of the majority of chunks, thus giving us only 11 chunks. The sizes of the new chunks were as follows: 4105, 5422, 4, 4, 7154, 14357, 10750, 5, 4491, 9423, and 2278. Three of these chunks (3, 4 and 8) have very small size, which made them suitable for discarding. Therefore, the interpolation process left us with only 8 chunks.

4. FILTERING OF THE TIDAL COMPONENTS

We performed discrete filtering of both upstream and downstream data using the Filter Design and Analysis (FDA) Tool in Matlab Signal Processing Toolbox, v.6.2 in order to remove the tidal periodic components from the data. The first idea was to create and use the infinite impulse response (IIR) filter (Proakis & Manolakis, 2006), because it can potentially meet the design specifications with lower order than the corresponding finite impulse response (FIR) filter, which would also result in shorter time to buffer the data. However, several attempts (using the Yule-Walker method, notch or elliptic filters) didn't achieve the expected results — the order was too high and the attenuation was less than specified (Pokrajac et al., 2007a). Hence, we designed the FIR filter. Since the spectrum of the data had peaks in two bands (see Fig. 5), two stopband filters were designed. Both of them had a passband ripple of 0.05, and the sampling frequency $fs = (1/30)$ min^{-1} = 0.556 mHz (Pokrajac et al., 2007a). In order to have a stopband attenuation of at least 20 dB in the $11 - 11.4$ µHz band, which corresponds to a 24.8 h period, the first created filter was of order 168. The attenuation of 40 dB in the $22.401 - 23.148$ µHz band (which corresponds to periods of 12 and 12.4 h) was achieved with the second filter of order $Nfilter = 354$. Here, more attenuation was needed due to the very high corresponding peak in the spectrum. In Figs. 10 and 11, magnitude responses of the first and the second filters are shown. The result of applying both filters on chunk 99 and downstream data is illustrated in Fig. 12. At the beginning of each chunk, we had to discard $N_{filter}-1$ data samples in order to perform filtering. This led to discarding less than 5% of the data. The standard deviation of the downstream data after the filtering was $std(y_{FIR}(t)) = 0.200$. Also, we tried the alternative approach by applying a moving average (MA) filter of length $Q = 25$, which corresponds to a period of 12.4 h. Standard deviation of the downstream data after the MA filter was $std(y_{MA}(t)) = 0.223$. The result of filtering the downstream data is shown in Fig. 13.

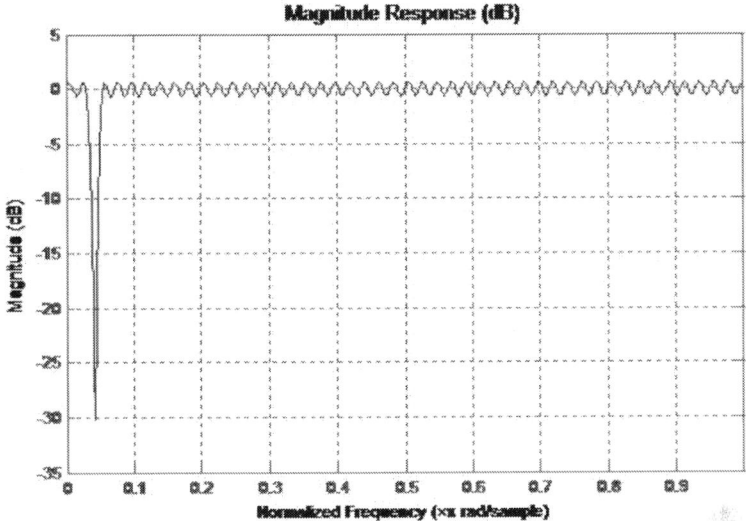

Figure 10. Magnitude response of the first filter.

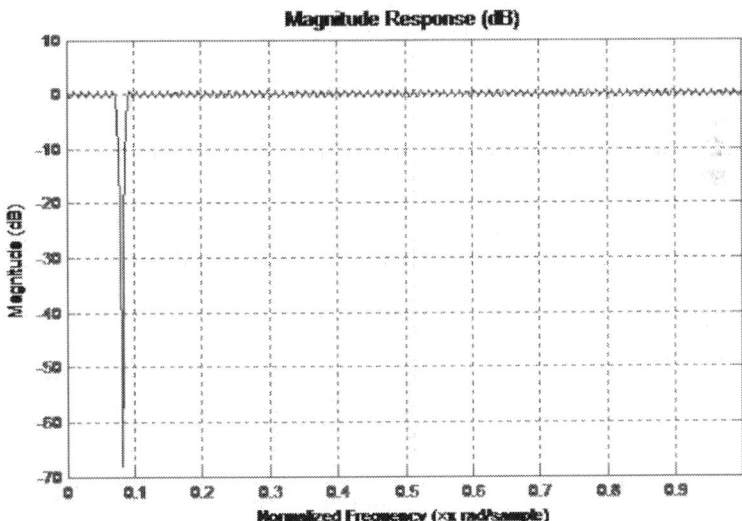

Figure 11. Magnitude response of the second filter.

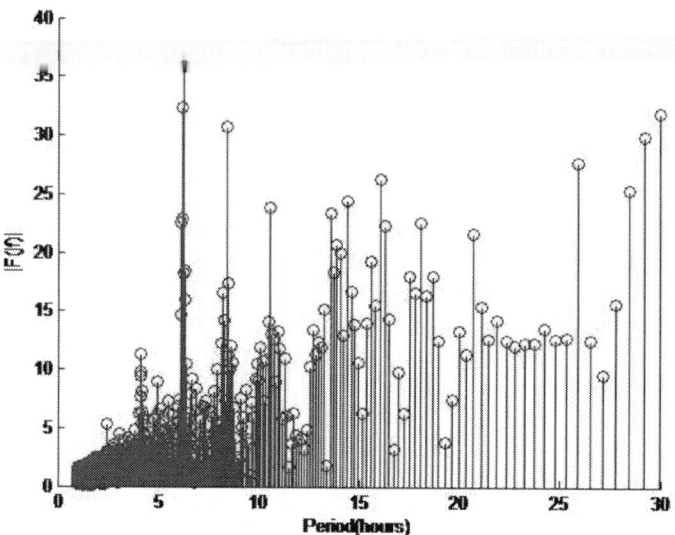

Figure 12. Spectrum after filtering (chunk 99 and downstream data).

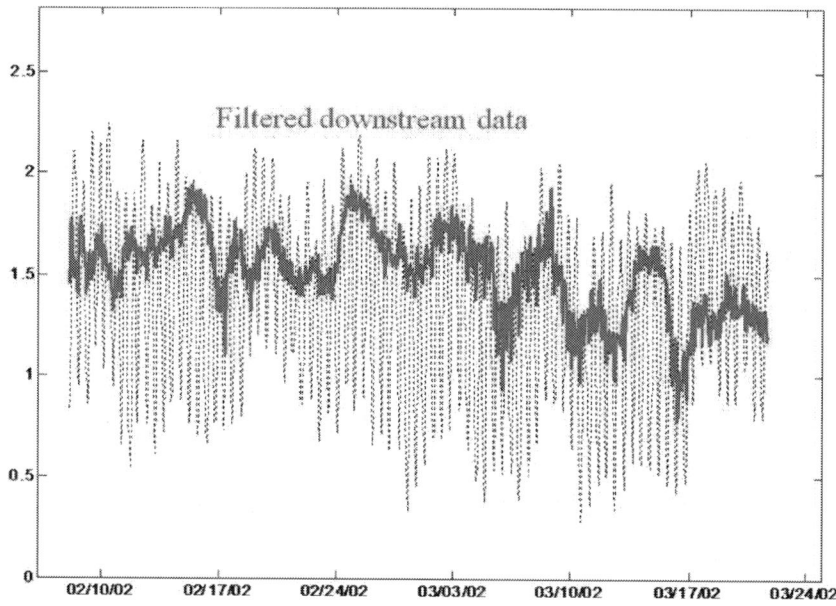

Figure 13. Filtered downstream data.

5. APPLICATION OF THE ADAPTIVE FILTERS

The downstream data y_t can be considered as a non-stationary function of the delayed upstream data x_t (see Fig. 14) (Pokrajac et al., 2007a, 2007b). It can be described as the discrete model $y_t = f_t(x_t, x_{t-T_s}, ..., x_{t-(L-1)T_s}) + r_t$, where L is the maximal delay of the model and r_t is the residual corresponding to the portion of the downstream data which cannot be explained by the upstream data.

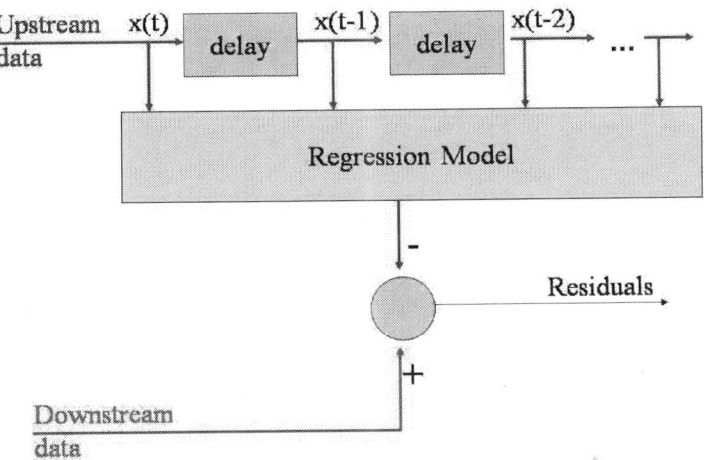

Figure 14. Removal of the upstream data influence.

If a function f_t is linear, the adaptive linear model can be represented as follows:

$$y_t = w_t^T x_t + r_t \qquad (2)$$

where $w_t = [w_{0,t} ... w_{L-1,t}]^T$ are coefficients and $x_t = [x_t ... x_{t-(L-1)T_s}]^T$ is the upstream data vector. A linear regression model could be obtained if the coefficients w are held constant (Devore, 2007):

$$y_t = w^T w_t + r_t \tag{?}$$

The coefficient of determination, R^2, is usually used to measure the accuracy of the model, (Devore, 2007). It is defined as a function of averaged squared residuals and the standard deviation of the response:

$$R^2 = 1 - \frac{\overline{\hat{r}_t^2}}{std(y_t)^2} \tag{4}$$

where the residuals are estimated with:

$$\hat{r}_t = y_t - w_t^T x_t \tag{5}$$

The updating of the coefficients wt in Eq. (2) is performed using the Widrow-Hoff least mean squares (LMS) algorithm (Widrow & Stearns, 1985):

$$w_{t+1} = w_t + 2\mu \hat{r}_t x_t \tag{6}$$

where μ represents the adjustable learning rate, and $\hat{r}_t$ is estimated using Eq. (5). In addition to the Widrow-Hoff LMS algorithm, we applied time notching by adjusting the coefficients only when all the time instants, $t,..., t-(L-1)T_s$, belonged to the same chunk of interpolated data (Pokrajac et al., 2007a, 2007b), see Fig. 15.

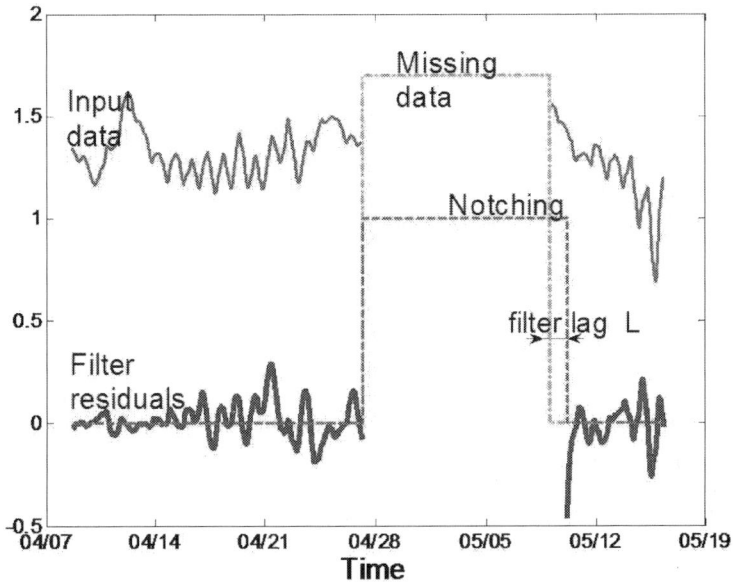

Figure 15. Time notching in adaptive filtering.

Using the linear regression given with Eq. (3) on the data $y_{MA}(t)$, which is processed by the MA filter, we were able to explain only 6% of the variance, i.e. $R^2 = 0.06$ for $L = 55$. In Table 1 are shown the results of obtained $std(\hat{r}(t))$, for different values of the learning rate and the filter delay, when the adaptive filter given with Eqs. (2), (5) and (6) is used. Useful models were obtained when $std(\hat{r}(t)) < std(y_{MA}(t)) = 0.200$, and are shown in the shaded boxes in Table 1. As can be seen, the best results were obtained for $L = 55$, $\mu = 0.015$, which yielded to $R^2 = 0.37$. Small values of the learning rate, combined with small filter length, lead to unsatisfactory results. On the other hand, the learning becomes unstable if the filter length and learning rates are getting large.

Table 1. *Std* of residuals for different values of learning rate μ and the filter length L, for MA filter. Useful models are in shaded boxes.

/L	30	35	40	45	50	55
0.01	0.226	0.213	0.201	0.19	0.187	0.18
0.015	0.204	0.19	0.174	0.164	0.16	0.157
0.02	0.183	0.17	0.159	1.846	4.80E+05	1.20E+13

When the designed FIR filter was used on the same data $y_{MA}(t)$, we received the results shown in Table 2. As can be seen, the results given in Table 1 are better. However, if l = 45 and μ = 0.015 (which varies to μ = 0.21), the best performance of the designed FIR filter is achieved.

Table 2. *Std* of residuals for different values of learning rate μ and the filter length L, for designed FIR filter. Useful models are in shaded boxes.

/L	20	30	35	40	45	50	55
0.01	0.25	0.23	0.21	0.21	0.19	0.19	0.17
0.015	0.23	0.2	0.2	0.18	0.17	24.9	5e4
0.02	0.21	0.19	0.47	2e6	5e16	3e30	1e48

Fig. 16 shows the residuals in time of the three useful adaptive filters, applied on data $y_{MA}(t)$ and processed by using the MA filter. The particular combination of the filter parameters was different, but the residuals showed similar behavior. Table 3 provides the time intervals when the relative residuals are larger than four standard deviations for the ada ptive filter with L = 55, μ = 1.5 e-2 (Pokrajac et al., 2007b). At the beginning of the learning process, the filter coefficients were not adapted fully, which caused the identified peaks. These peaks corresponded to observations from February 2002. In addition, there are two other peaks, corresponding to Sep 4, 2002, and Oct 25, 2004, which can be explained by a transient behavior after the notching interval.

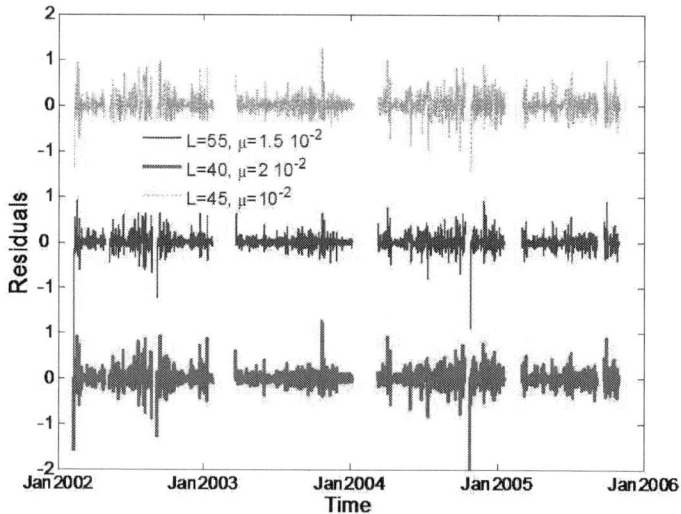

Figure 16. Residuals of the adaptive filters applied on $yMA(t)$ data using three different combinations of the learning rate and the filter length.

Table 3. Identified intervals of large residuals (adaptive filtering on $yMA(t)$ data, $L = 55$, $\mu = 1.5\,e\text{-}2$

Year	Begin		End		Note
	Date	Time	Date	Time	
2002	8-Feb	23:00	10-Feb	0:00	Learning
	17-Feb	6:30	18-Feb	1:30	Learning
	20-Jul	0:30	20-Jul	9:00	
	7-Aug	1:00	7-Aug	3:30	
	7-Aug	16:30	7-Aug	17:30	
	21-Aug	14:30	21-Aug	21:30	
	4-Sep	16:00	4-Sep	19:30	Transient
	11-Sep	15:00	12-Sep	11:00	
	21-Dec	9:30	21-Dec	10:30	
2003	9-Jan	2:00	9-Jan	4:00	
	22-Oct	12:00	24-Oct	9:00	

Year	Begin		End		Note
	Date	Time	Date	Time	
2004	1-Apr	15:30	2-Apr	18:30	
	9-Apr	17:00	9-Apr	18:00	
	29-May	12:30	29-May	19:00	
	12-Jul	20:30	13-Jul	2:00	
	19-Sep	0:00	19-Sep	6:00	
	29-Sep	20:30	30-Sep	6:30	
	5-Oct	15:30	5-Oct	20:30	
	25-Oct	3:00	25-Oct	6:00	Transient
	26-Nov	9:00	26-Nov	15:00	
2005	27-Sep	20:00	28-Sep	4:30	

6. CONCLUSION

We have described the application of the adaptive filtering for analyzing river hydrographic data. When determining the portion of the downstream data that is not influenced by the upstream data, the numerical results show that adaptive filtering is superior to linear regression.

7. ACKNOWLEDGEMENTS

This work was partially supported by the US Department of Commerce (award #NA06OAR4810164), NOAA (#NA06OAR4810164), NIH (#2P20RR016472-04), DoD/DoA (#45395-MA-ISP, #54412-CI-ISP, W81XWH-09-1-0062), and NSF (#0320991, CREST grant #HRD-0630388, #HRD-0310163).

REFERENCES

1. P. Adam, 2002 Saltmarshes in a time of change. Environmental Conservation, 29 1 3961

2. Delaware Department of Natural Resources and Environmental Control (DNREC)(2002). Technical Background Report Silver Lake Watershed, 81 Dover, Delaware, USA

3. J. Devore, 2007 Probability and Statistics for Engineering and the Sciences (7th Edition), Duxbury Pr., 0-49555-744-7

4. Fondriest Environmental Inc, http://www.fondriest.com

5. C. Hull, J. Titus, 1986 Greenhouse Effect, Sea Level Rise, and Salinity in the Delaware Estuary, EPA report 230-05-86-010

6. M. Kennish, 2002 Environmental Threats and Envorinmental Future of Esuaries. Environmental Conservation, 29 1 78107

7. J. Kraft, T. Hi-ll, M. Khalequzzaman, 1992 Geologic and Human Factors in the Decline of the Tidal Salt Marsh Lithosome: the Delaware Estuary and Atlantic Coastal Zone. Sedimentary Geology, 80 232246

8. M. Mensinger, 2005 SWMP Metadata, In: National Estuarine Research Reserve Centralized Data Management Office,http://cdmo.baruch.sc.edu /QueryPages/data_summary_ progress.cfm

9. W. Mitsch, J. Gosselink, 2000 Wetlands, John Wiley and Sons Inc., 047129232 New York

10. R. Morris, I. Reach, M. Duffy, T. Collings, R. Leafe, 2004 On the Loss of Saltmarshes in South-East England and the Relationship with Nereis Diversicolor. Journal of Applied Ecology, 41 787791

11. D. Pokrajac, N. Reljin, M. Reiter, S. Stotts, R. Scarborough, 2007 Signal Processing of St. Jones River, Delaware Water Level Data, Proceedings of ETRAN 2007, Herceg Novi, Montenegro, Jun 2007

12. D. Pokrajac, N. Reljin, M. Reiter, S. Stotts, R. Scarborough, J. Nikolic, 2007 Adaptive Filters for Water Level Data Processing, Proceedings of TELSIKS 2007, Nis, Serbia, September 2007

13. I. C. Potter, R. Manning, N. Lonergan, 1991 Size, Movements, Distribution and Gonadal Stage of the Western King Prawn in a Temperate Estuary and Local Marine Waters. Journal of Zoology, 223 419445

14. J. Proakis, D. Manolakis, 2006 Digital Signal Processing (4th Edition), Prentice Hall, 0-13187-374-1

15. J. Rooth, J. Stevenson, 2000 Sediment Deposition Patterns in Phragmites Aurstalis Communities: Implications for Coastal Areas Threatened by Rising Sea-level. Wetlands Ecology and Management, 8 173183

16. S. Temmerman, G. Govers, S. Wartel, P. Meire, 2004 Modeling Estuarine Variations in Tidal Marsh Sedimentation: Response to Changing Sea Level and Suspended Sediment Concentrations. Marine Geology, 212 119

17. B. Widrow, S. Stearns, 1985 Adaptive Signal Processing, Prentice Hall, 0-13004-029-0

18. M. Zhang, S. Ustin, E. Rejmankova, E. Sanderson, 1997 Monitoring Pacific Coast Salt Marshes Using Remote Sensing. Ecological Applications, 7 3 10291053

19. Y. Zharikov, G. Skilleter, N. Loneragan, T. Tarant, B. Cameron, 2005 Mapping and Characterizing Subtropical Estuarine Landscapes using Aerial Photography and GIS for Potential Application in Wildlife Conservation and Management. Biological Conservation, 125 87100

20. Y. Zharikov, G. Skilleter, 2004 Potential Interactions between Humans and non breeding Shorebirds on a Subtropical Intertidal Flat. Australian Ecology, 29 647660

Chapter 9

Anti-Multipath Filter with Multiple Correlators in GNSS Receviers

Chung-Liang Chang

[1] *Department of Biomechatronics EngineeringNational Pingtung University of Science and TechnologyPingtung County, Taiwan*

1. INTRODUCTION

The positioning tecnique of global navigation satellite systm (GNSS) has become mature and also been applied to a variety of navigation vehicles, whether it be the application to ground vehicle ot aircraft. Nevertheless, the precision of GNSS is susceptible to intentional or unintentional factors such as interference or jammer, etc. The influences range from minor effect like the positioning precision of satellite signal to significant impact like the misleading information to users or malfunction of receivers. The ionosphere or troposphere in environment or the noise in receiver itself are the source of positioning error when satellite passes through ionosphere or troposphere, the change of media results in the delay of wave transmission rate and yields error. The adoption of dual frequency receiver can decrease error but it presents no significant improvement in terms of the error generated by multipath.

The effect of multipath is because the satellite signal is reflected or diffracted by obstacle prior to its receiption by antenna. Most of the time, it results in the decrease of signal propagation power and delay of time. In 1973, Hagerman employed conventional code tracking to analyze the effect of multipath on the coarse/acquire (C/A) code in carrier L1 using one chip of early-late spacing. He also estimated that under different delay, phase and signal magnitude, the effect may result in 70–80 m tracking error [Hagerman, 1973]. With the growing application of global positioning system (GPS), many researches investigating multipath effect have been proposed to effectively reduce its impact and have provided various estimation algorithms for implementation in hardware.

The most effctive solution for multipath offset is the location of antenna. Assume the antenna is placed above the highest source reflection, the reflected signal will not be received. In antenna design, it can reduce the gain of received signal coming through lower elevation. Generally, the receiver will setup up the minimum elevation capable of receiving satellite signal. The design of choke ring antenna is used to mitigate multipath. The choke ring antenna circles the antenna with vertical concentric rings, whose function is to reduce the gain of received reflected signal. However, such a function is strongly related to the location of antenna.

In additioon, the change of structure in internal correlator design of receiver is also a solution for multipath. The conventional GPS receiver typically adopts one chip early-late spacing of correlator. The use of narrow correlator to reduce chip spacing can effectively mitigate multipath and noise, which cuts down the error of 70–80 m to 8–10 m (van Dierendonck et al., 1992). Note that the use of narrow correlator technique in coherent discriminator may lead to the lock failure in code delay locked loop without the cooperation of phase locked loop (PLL).

The strobe correlator and edge correlator are both solutions for multipath mitigation (Garin et al., 1996). The strobe correlator is implemented using two different narrow correlator discriminators. The strobe correlator and edge correlator developed by Ashtech only provide code correlation for C/A code. The enhanced strobe correlator (Garin and Rousseau, 1997) offers carrier phase correction and code correction for C/A code. With the additional carrier phase correction in terms of multipath its real-time dynamic processing outperforms previous methods. Note that the narrow correlator and strobe correlator do not encompass carrier phase correction. Thus, their sensitivity approaches that of conventional correlator.

Another discriminator design is early 1/ early 2 (E1/E2) correlator (Mattos, 1996; van Dierendonck and Braasch, 1997). The method utillzes part of correlation coefficients not subject to multipath effect for multipath mitigation. That is, it employs two correlators with the spacing and location at the front end of correlation function. However, this method is a choice between noise mitigation and multipath mitigation.

The multipath estimation method initial estimates multipath signal and then subtracts it from received signal so that the signal approaches direct signal. Literature review that resembles this algorithm are MEDLL, MET. (van Nee, 1992; van Nee et al., 1994), which utilize maximum likelihood estimation technique and recursive least square method to estimate the magnitude, delay, phase and erase it from received signal. Though the above estimation methods can not completely eliminate multipath signal, they present significant improvment in terms of multipath delay within certain range.

Nevertheless, these techniques have difficulties in mitigating short-delay multipath signals (less than 0.1 PN code chip or approximately 30 m). Scholars have proposed methods on short-delay multipath mitigation (Sleewaegen et al., 2001; Stone and Chansarkar, 2004). However, these techniques still have drawbacks. The method proposed by Sleewaegen requires a scaling factor, depending on multipath environment, to link the signal amplitude with the range error. The method proposed by Stone and Chansarkar is to estimate the pseudorange error on the basis of a statistical model, which requires large numbers of collected data. Consequently, the performances of these two methods are significantly influenced by multipath environment.

The author has proposed an adaptive filter in 2008 (Chang and Juang, 2008), which adopts five tap-delay to effectively mitigate short-delay multipath. Though this method is efficient in short-delay multipath mitigation, it does not guarantee that the receiver will not receive multipath signal at different time delay under variable environment. Moreover, the correlator technique of coventional receiver is not quite capable of accurately describing the data distribution of correlated signal, which results in longer period of time to estimate multipath parameter. Thus, this paper utilizes multi-correlator technique in combination with proposed method to mitigate the mystical multipath signal. Simulation results show that the multi-correlator technique can clearly present the output distribution of correlator, make adaptive filter rapidly estimate multipath parameter and cope with multipath signal at different time delay.

2. METHODOLOGY

2.1. Multipath Overview

Multipath effect is caused by the reflection of satellite siganl by obstacles when the receiver receives the reflected signal, it leads to positioning error and the lock failure of signal for receiver, which renders positioning funciton void. In GPS, the desired signal consist of only the direct path signal. All other signals distort the desired signal and result in ranging measurement errors. To understand the effect of multipath in measurement process, let's consider the heart of the GPS code tracking loop. The pseudorange measurement originates from a locally generated pseudorandom noise (PRN) code which is kept phase-locked to the received code. The discriminator is formed based on the difference between early correlator output and late correlator output. The output of the discriminator is fed back to the local code generator to keep synchronism between the local code and incoming code. This generatess the so-called delay-locked loop (DLL). When multipath is present, the incoming code, correlation function and discriminator functions are distorted. Analytically, the direct and multipath components may be conducted separately. Note that for the direct-path case, the discriminator function that passes through zero when the code-tracking error (local-code delay) is zero. This is the ideal case. However, when multipath is present, the distorted function has a zero-crossing at non-zero code tracking error. Fig. 1 demonstrates the tracking errors of the early-late discriminator output due to multipath in the DLL. The tracking errors result from distortion of the correlation function with the received IF signal. In the direct-path case, the ideal case is when the discriminator function passes through zero while the code tracking error is zero. However, with the presence of multipath, the distorted function has a zero-crossing at a non-zero code tracking error. With the direct signal, when the relative multipath phase is 0 radians, the multipath component is 'in-phase'. With pi radians, the multipath component is 'out-of phase'.

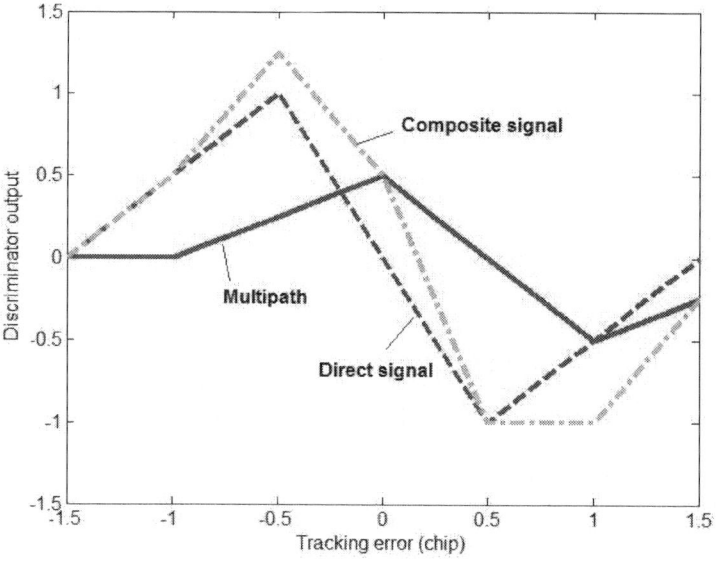

Figure 1. Composite distorted of early-late discriminator.

Thus, pseudorange multipath analysis encompasses simulation of direct and indirect path signals and determination of zero-crossing of distorted discrimintator function. There are three multipath parameters to consider: strength, delay and phase. The absolute value of each parameter is irrelevant. The upper and lower bounds of the multipath error can be determined, for a given multipath-to-direct ratio, by fixing the relative multipath phase at 0 and pi radians, respectively, and varying the relative multipath delay. At each delay point, the distorted discriminator curve is determined and the resulting zero-crossing point and pseudorange error are calculated. The result of an example is presented in Fig. 2, which illustrates result of the theoretical multipath error envelope versus the multipath delay. The code autocorrelation sidelobes have been ignored. This simulation is offered in the case of 24 MHz bandwidth receiver filter, 1-chip, 0.5-chip, and 0.2-chip early-late (E-L) spacing and unaltered multipath amplitude. A conventional GPS receiver adopts a delay-lock loop with a 1-chip spacing between early and late correlators. The smaller E-L spacing is regarded as narrow-correlator architecture. Narrow-correlator receivers typically utilize spacings in the range of 0.05 to 0.2 PRN chips.

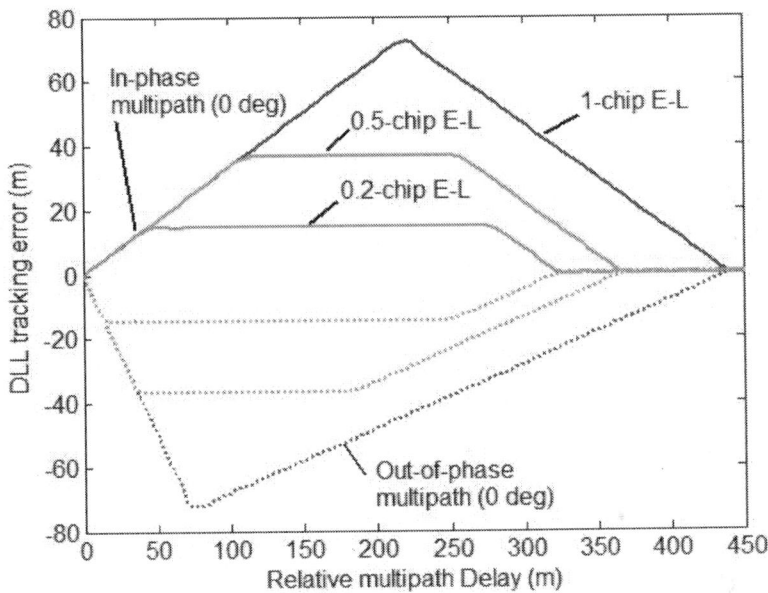

Figure 2. Multipath error envelope for a conventional, 1-chip early-to-late (E-L), 0.5-chip E-L, and 0.2-chip E-L DLL receiver; Multipath component is half the strength of direct signal.

2.2. Signal Model

A GPS receiver may receive a number of reflected signals and direct signal from the satellite. The error source of GPS consist of ionosphere delay, troposphere delay, receiver noise and multipath effect. Except for multipath, the other errors can be significantly decreased through advanced prediction and differential correction method. It is hard to depict the statistical model of the received signal in the presence of multipath. However, many hypotheses can still be proposed. One hypothesis describes that the multipath signals are delayed with respect to direct GPS signal. Thus, let's consider only these reflected signals with a delay of less than one chip. This is because signals with a code delay larger than one chip are uncorrelated with the direct signals. Otherwise, the multipath signal is assumed to have the lower power than the direct one. The composite baseband signal, ignoring the navigation data bit, is given by

$$z[n] = \sum_{k=0}^{M} \alpha_k p(nT_s\text{-}\tau_k) \exp(-j(\omega nT_s + \theta_k)) + \nu[n] \tag{1}$$

Where α_k, θ_k and τ_k denote amplitude, carrier phase, and code delay of k-th delayed signal. M represents the number of multipath component. $p(\cdot)$ indicates spread-spectrum code. Ω denotes the IF angular frequency. The notation $z[n] = z(nT_s)$ is employed to denote a digital sequence sampled at the frequency $f_s = 1/T_s$ where T_s indicates period of sampling and nn is the discrete time index. The 0-th delayed signal corresponds to the direct signal. $\nu[n]$ is modeled as white Gaussian noise distribution. The positioning error caused by the reception of multipath and direct signal is not only associated with the hardware design of receiver but also the detection algorithm. The literature review has provided several solutions for multipath effect. The following chapter will describe the proposed algorithm to counteract multipath.

2.3. Multiple Correlator Concept

The design of multi-correlator is seldom implemented due to the consideration of processing speed of hardware and cost. Owing to the promotion of hardware speed, decrease of cost and emergence of software wireless, the application of multi-correlator technique to receiver has become more prevalent. In fact, the strobe correlator described above is one of multi-correlator technique, which utilizes the linear combination of two correlators as discriminator output and adjusts chip spacing to track signal. Multi-correlator technique can depict the signal distribution after correlation process. In other words, this technique can present the process of correlation output in detail. Fig. 3 demonstrates the correlation output using 1 and 32 correlators, respectively. This Figure illustrates that the multipath component can not be apparent if it adopts one set of correlator (early, prompt, and late). On the contrary, the 32 sets of correlator can better present the distribution of correlation output. Assume there are five correlators and the correlation of received signal is known. The linear combination of the five correlators can constitute received signal, which is expressed as:

$$v_j = \sum_{i=1}^{Q} r_{ji} u_i \qquad (2)$$

V denotes each measurement value of correlator, Q indicates the number of correlator, r is corresponding correlation value and u is the scaled value of correlation center itself. Take the five correlators as example. Assume five correlators are located at -0.5, -0.25, 0, 0.25, 0.5, respectively. The combination of five correlators can be employed to accomplish the measurement value of each correlator. Equation (1) is rewritten as follows:

$$v = R\Lambda$$

$$\Rightarrow
\begin{bmatrix}
0.5 \\
0.75 \\
1 \\
0.75 \\
0.5
\end{bmatrix}
=
\begin{bmatrix}
1 & 0.75 & 0.5 & 0.25 & 0 \\
0.75 & 1 & 0.75 & 0.5 & 0.25 \\
0.5 & 0.75 & 1 & 0.75 & 0.5 \\
0.25 & 0.5 & 0.75 & 1 & 0.75 \\
0 & 0.25 & 0.5 & 0.75 & 1
\end{bmatrix}
\cdot
\begin{bmatrix}
0 \\
0 \\
1 \\
0 \\
0
\end{bmatrix}
\qquad (3)$$

The makeup of v is consitituted by third correlation (location as 0) because the Λ of the rest correlators is 0. The makeup of R is based on the location of selected correlator. With the v and R matrix known apriori, the magnitude of the signal in terms of the distribution set up by correlator can be known based on $\Lambda = R^{-1}v$. The more the correlators, the clearer the distribution of the signal.

It is known that the muti-correlator can depict the makeup of signal. Thus, we will see if multi-correlator can estimate direct signal with the direct signal plus multipath signal. Assume the multipath delay as 0.25 chip, signal magnitude as 0.5 and five correlators are shown as Fig. 4. Based on $\Lambda = R^{-1}v$, the distribution of signal is known. Apparently, a correlation value exists between third and fourth correlator and the Λ of fourth correlator is lower. Using the negative correlation value form fourth correlator, we can elminate multipath. Fig. 5 illustrates the multipath mitigation when the location of time delay is at the location of set multi-correlator.

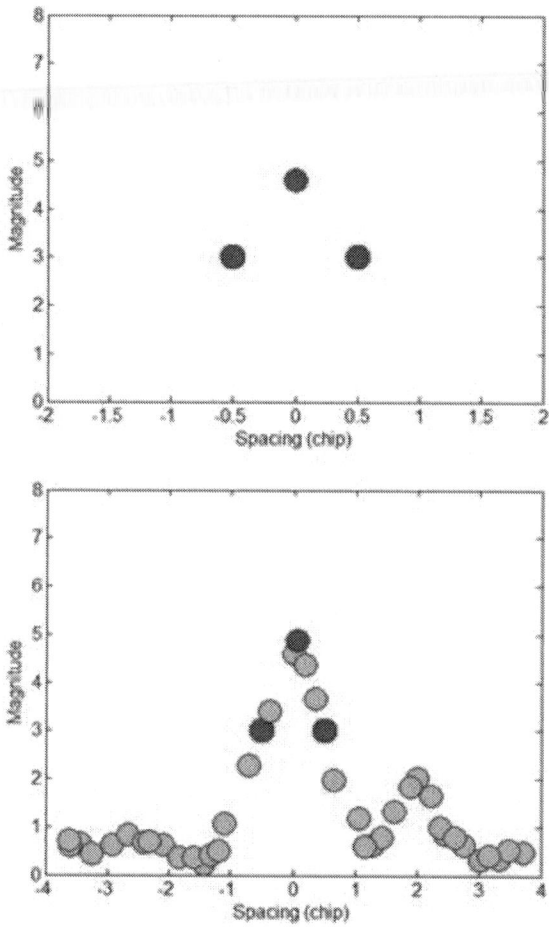

Figure 3. Comparision of single correlator and multi-correlator.

$$\Lambda = \mathbf{R}^{-1}\mathbf{v}$$

$$= \begin{bmatrix} 1 & 0.75 & 0.5 & 0.25 & 0 \\ 0.75 & 1 & 0.75 & 0.5 & 0.25 \\ 0.5 & 0.75 & 1 & 0.75 & 0.5 \\ 0.25 & 0.5 & 0.75 & 1 & 0.75 \\ 0 & 0.25 & 0.5 & 0.75 & 1 \end{bmatrix}^{-1} \begin{bmatrix} 0.625 \\ 1 \\ 1.375 \\ 1.25 \\ 0.875 \end{bmatrix} \approx \begin{bmatrix} 0 \\ 0 \\ 1 \\ 0.5 \\ 0 \end{bmatrix} \quad (4)$$

When the multipath delay is not at the set correlator, the calculated value after the above deduction approximates direct signal with little gap. Fig. 6 demonstrates the scenario when the location of multipath time delay is not at the location of set correlator.

$$\Lambda = \mathbf{R}^{-1}\mathbf{v}$$

$$= \begin{bmatrix} 1 & 0.75 & 0.5 & 0.25 & 0 \\ 0.75 & 1 & 0.75 & 0.5 & 0.25 \\ 0.5 & 0.75 & 1 & 0.75 & 0.5 \\ 0.25 & 0.5 & 0.75 & 1 & 0.75 \\ 0 & 0.25 & 0.5 & 0.75 & 1 \end{bmatrix}^{-1} \begin{bmatrix} 0.55 \\ 0.925 \\ 1.3 \\ 1.175 \\ 0.95 \end{bmatrix} \approx \begin{bmatrix} 0 \\ 0 \\ 1 \\ 0.2 \\ 0.3 \end{bmatrix} \qquad (5)$$

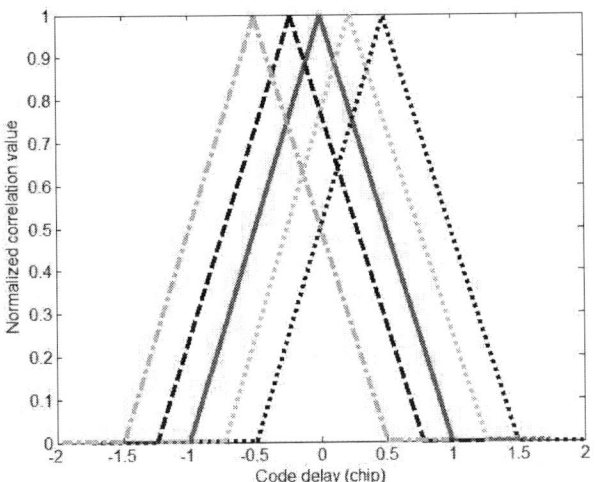

Figure 4. Distribution of five correlators.

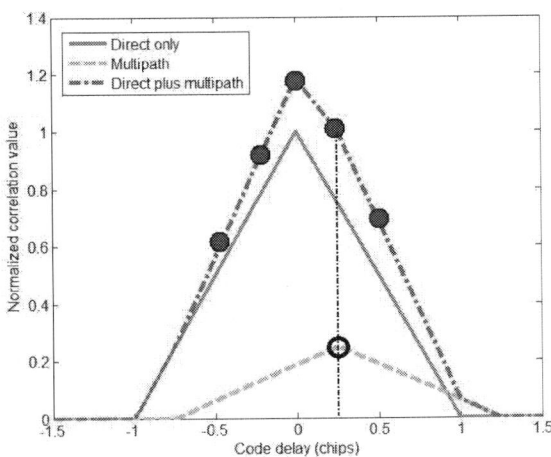

Figure 5. Multipath delay is at the set correlator.

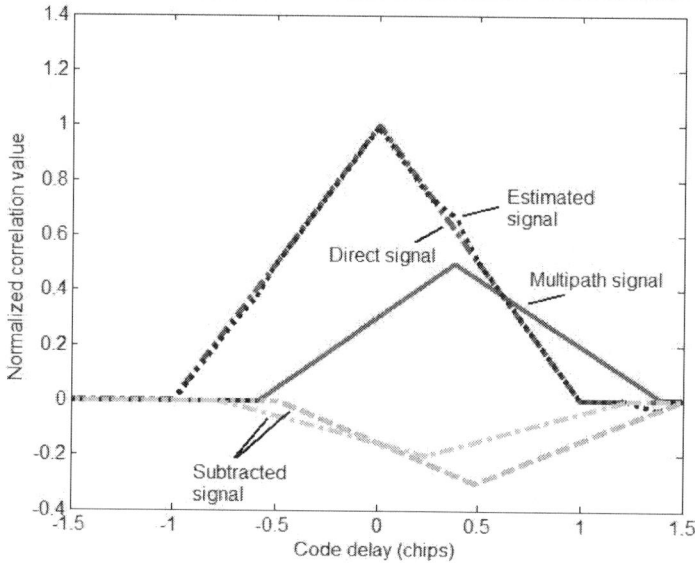

Figure 6. Multipath delay is not at the set correlator.

2.4. Anti-Multipath Filter With Multiple Correlator

The previous chapter has clearly presented the advantage of multi-correlator method and its operation process. This chapter will elaborate how to constitute an anti-multipath filter based on multi-correlator.Fig. 5 shows the block diagram of the multipath mitigation system. The received signal is processed in a RF filter, then downconverted and sampled to a digital IF signal.

The tracking module consists of multiple correlator, code/carrier generator, discriminator and filter. The purpose of this module is to acquire accurate code phase and the carrier phase from PLL and DLL. The multipath estimator is used to estimate the correlation parameter of multipath, on the basis of the adaptive filter by employing duplicated signal and digital IF signal. Fig. 7 demonstrates that the estimated signal parameters are sent to the correlation decomposer and the correlation value of multipath signal is determined in the multipath cancellation area.

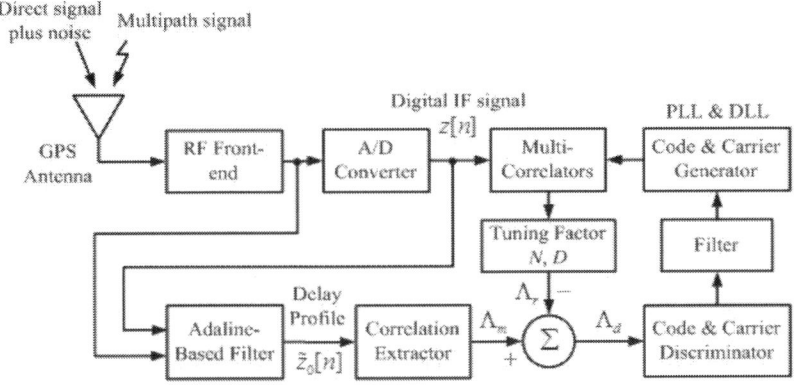

Figure 7. Multipath Mitigation System Block Diagram.

The estimated delayed signal is recreated at the Adaline-based filter and is subtracted from the correlation value of the received signal. The process of multi-correlators, multipath estimator, correlation value decomposer, and multipath cancellation will be elaborated in the following subsections.

2.4.1. Multi-Correlators Techniques

The concept of multi-correlator and the process of this method have been detailed in previous chapter. What we consider for the time being is that initial point of code delay of received signal and the local replica is not identical and multipath does not take place at the set correlator. Thus, paralell shift method is utilized to change the element of RR matrix, such as shift the correlator location. Based on the simulation, assume the code shift of received signal as 0.3 chip and multipath delay as 0.5 chip. Using the above method, we add two variable as shift times N and shift range D, respectively. The purpose is to acquire the received direct signal and counteract multipath. The circle in red in the following Figure are the code shift of direct signal, the shift times and range of correlator. The following will present the process. Fig. 8 (a) denotes the correlator output without shift operation. The color green is direct signal, the dark brown is multipath and the brown denotes composite signal. Afterward, the correlator is shifted 0.1 chip (N=1 and D=0.1), and Fig. 8 (b) is derived. However, this Figure reveals that the performance does not meet our expectation. Fig 8(c) illustrates that after shift 0.3 chip (N=1 and D=0.3), the brown siganl and green direct signal almost overlaps. The program is to simulate the location of set correlator in order to acquire v. Through the variation of R, multipath is mitigated. The result presents that acquired signal of correlator output using parallel shift method (shift 0.3 chip) is a more efficient strategy in multipath mitigation as opposed to shift 0.1 chip without shift.

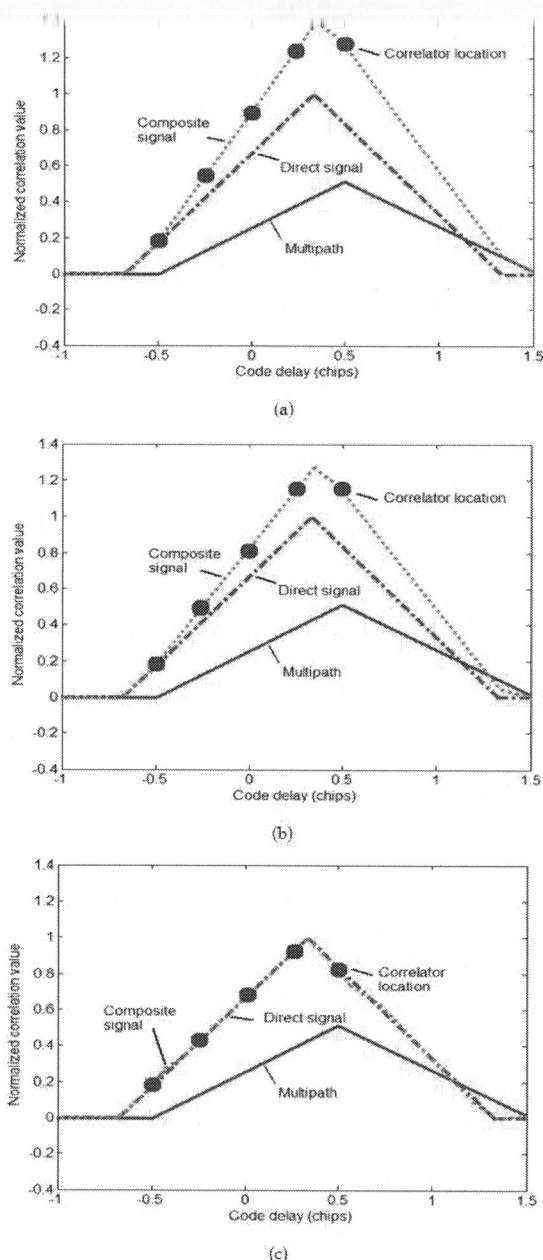

Figure 8. Multi-correlators technique simulation results. (a) N=0, D=0.1 (b) N=1, D=0.1 (c) N=3, D=0.1.

2.4.2. Adaline-Based Filter

The function of a multipath estimator is to estimate the multipath delay using Adaline-based filter, shown in Fig. 9. It adopts the tap-delay line with an Adaline network (Widrow and Hoff, 1960) to constitute this structure without a non-linear element. An adaptive algorithm such as the LMS algorithm or the Back-Propagation (BP) learning algorithm is often employed to adjust the weights of the Adaline so that it responds accurately to as many patterns as possible in a training set. It is the simplest and most intelligent self-learning system which adapts itself to achieve an optimal solution (Rumelhart, D. E. et al, 1986). In this paper, the BP with an adaptive learning rate algorithm serves as a substitute for the LMS algorithm so as to prevent inherent limitations in LMS and to improve filter convergence rate (Schalkoff, R. J., 1997). The multipath estimator offers the multipath delay profile. Suppose the estimated digital IF signal is given by:

$$\widetilde{z}\,[n] = \sum_{k=0}^{\widetilde{M}} \widetilde{\alpha}_k p(nT_s - \widetilde{\tau}_k) \exp(-j(\omega nT_s + \widetilde{\theta}_k)) + \nu[n] \tag{6}$$

Where the parameter with the symbol "~" denotes the estimated parameter. A reference signal is a replica of code and carrier deriving from the output of DLL and PLL, which is shown as:

$$c_k[n] = p(nT_s - kd_\tau - \varepsilon_\tau) \exp(-j(\omega nT_s - \varepsilon_\theta))(k = 0, \ldots, K) \tag{7}$$

where ε_τ and ε_θ denote the measured group delay and carrier phase consisting of multipath error. d_τ indicates sample period of delay of the multipath signals and $Kd_\tau Kd_\tau$ denotes the maximum delay of multipath signals. It is difficult to determine the parameters directly without any assumption about multipath signals. Thus, (6) is adopted in estimation process and modified by using reference signal and replacing $\widetilde{M}$ ~ with $\widetilde{K}$ where the output signal of the filter is expressed as:

$$\widetilde{z}\,[n] = \sum_{k=0}^{\widetilde{K}} w_k c_k[n] + w_b c_b + \widetilde{\nu}[n] \tag{8}$$

where $w_k = \widetilde{\alpha}_k \exp(j\widetilde{\theta}_k)$ represents the adjustable weight. The filter weight is employed to minimize the cost function, called squared error energy function and defined by using Equation (1) and (6):

$$L[n] = E\left\{ (z[n] - \widetilde{z}[n])(z[n] - \widetilde{z}[n])^H \right\} \tag{9}$$

The filter minimizing the cost function is chosen by its tap weights to be the optimal solution to the normal equation (Haykin, S., 1986).

$$w_k^{\text{opt}} = C_k g_k^{-1}, \tag{10}$$

where c_k denotes the autocorrelation, $E\left\{ x_l[n] x_k^H[n] \right\}$, of two reference signals ($c_l[n]$ and $c_k[n]$). g is the cross-correlation, $E\left\{ z[n] c_i^H[n] \right\}$, of the digital IF signal $z[n]$ and reference signal $zl[n]$. Where E{·} indicates an expectation operator. The filter solves (10) recursively using the BP with adaptive learning rate algorithm. This learning rule performs a gradient descent on the energy function to derive a minimum:

$$w_k[n] = w_k[n\text{-}1] + \mu\delta[n\text{-}1] c_k[n\text{-}1],$$
$$w_b[n] = w_b[n\text{-}1] + \mu\delta[n\text{-}1] c_b \tag{11}$$

where $\delta[n]$ denotes the output layer error term. $\tilde{\alpha}_k$, $\tilde{\theta}_k$, and $\tilde{\tau}_k$ are estimated as the absolute value of weight $|w_k|$, the phase angle of weight arg(w_k) and the value of delay element kd_τ, respectively. The bias weight w_b connected to a constant input c_b=+1, effectively controls the input signal level of the filter. Note that the digital IF signal given in (1) is adopted as the desired signal and the output of DLL and PLL serves as the filter input signal. The reference signal is determined by (7) which generates the output of each delay element. Therefore, the estimated delay parameters from the filter weights and the delay element can be derived on condition that the learning algorithm has converged. The learning rate coefficient $\mu\mu$ determines stability and convergence rate and a BP trained reference signal is adopted to obtain the minimum of (9) (Widrow 1986; Jacobs 1988). Suppose the learning rate is too large, the search path will oscillate about the desired path and converge more slowly than a direct descent. Nevertheless, the descent will progress in small steps if the learning rate is too small. It will greatly significantly increases the total time to convergence. Consequently, an adaptive coefficient where the value of μ is a function of the error derivation is adopted as the solution (Schalkoff, R. J., 1997).

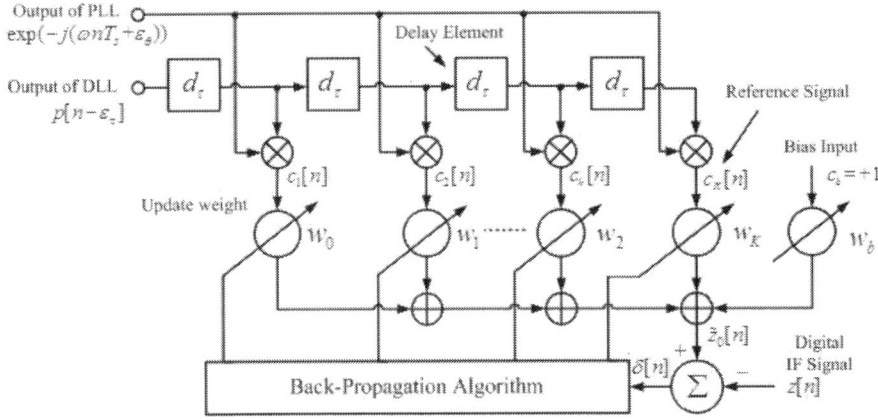

Figure 9. Structure of the Adaline-based filter used in the multipath estimator.

2.4.3. Correlation Extractor

After the use of adaptive filter, the estimated parameters can be obtained and the correlation decomposer divides the estimated parameters into multipath and direct signal. Besides, the autocorrelation function of multipath signals is subtracted from analog-to-digital (A/D) converter output of the received signal. In the decomposer process, it is assumed that the values of the first peak amplitude tap weight are the direct signal and the remainders are multipath signals. Fig. 5 presents an example where the direct signals refers to the first peak k=l and the multipath signal amplitude denotes the remnants /k≤K. Suppose that the multipath channel has a decreasing power delay profile. The multipath signal parameter is adopted to calculate the correlation value. The correlation equation of estimated multipath signals with amplitude $\tilde{\alpha}_k$, delay $\tilde{\tau}_k$ and carrier phase $\tilde{\theta}_k$ is written as:

$$\Lambda_k(\tau) = \tilde{A}_k C(\tau - \tilde{\tau}_k)\cos(\tilde{\theta}_l - \tilde{\theta}_k), \tag{12}$$

where C(τ) denotes the autocorrelation function of the GPS PRN (Pseudo Random Noise) code signal.

$$\mathrm{E}\{p[n]p[n-\tau]\} \tag{13}$$

Hence, the entire correlation value of the estimated multipath signal Λk(τ)Λk(τ) is given by:

$$\Lambda_{\mathrm{m}}(\tau) = \sum_{k=l+1}^{\overline{K}} \Lambda_k(\tau)$$

(14)

2.4.4. Multipath Removal

The entire correlation values of multipath signal $\Lambda_m\Lambda_m$ are subtracted from the correlation value of received signal $\Lambda_r\Lambda_r$ and the output of correlation value $\Lambda_d\Lambda_d$ is expressed as:

$$\Lambda_{\mathrm{d}}(\tau) = \Lambda_r(\tau) - \Lambda_{\mathrm{m}}(\tau)$$

(15)

In (12), the estimated correlation $\Lambda_r\Lambda_r$ of direct signal can be acquired using multi-correlator technique. Such a technique has been detailed in chapter 2.3.1. Multi-correlator technique can effectively estimate the correlation of direct signal and counteract multipath simultanously. It can promote the convergence speed of Adaline-based filter.

The tracking error takes place in DLL and PLL due to the multipath effect. The effect primarily results from distortion of the correlation function receiving the IF signal, shown in Fig. 8(a), which illustrates the normalized correlation function with multipath effect. Fig. 8(a) presents that the symmetry is lost and the propagation delay is difficult to estimate. Thus, the range measurement accuracy is diminished. Nevertheless, the use of a subtractive method offers multipath mitigation in the tracking loop and the output $\Lambda_m(\tau)$ enables the tracking loops to track direct signal accurately.

The above processes: the estimate process, the correlation extractor and the cancellation method can counteract the multipath effects regarding the autocorrelation function of the received signal, since the tracking errors in DLL and PLL are not completely eliminated. Provided that the reference signal acquires the multipath error, the estimated parameters do not present accurately that of the real multipath. So as to obtain the ideal estimated parameters, the BP learning process is recursively employed. The use of multi-correlator technique can speed up BP learning process and enhance its performance.

3. PERFORMANCE ANALYSIS AND SIMULATION RESULTS

In this section, computer simulations are performed to evaluate the performance of proposed method. To compare with other published methods in performance, the multipath tracking error envelopes in code and carrier phase for a multipath signal amplitude of half the LOS amplitude is denoted as $\alpha_0=1.0$ and $\alpha_1=0.5$. A GPS multipath model includes one direct signal and one delayed signal. Suppose that a high post signal to noise ratio (SNR) of 10 dB is located in this model. Simulation results are demonstrated in infinite bandwidth situation.

3.1. Simulation Parameter

The digital IF frequency of a GPS signal is $\omega/2\pi=1.25$ MHz and the sampling rate is 5 MHz. The delay chip of the multipath signal varies from 0 to 1.5 chips with the phase of 0 and π radians with regard to the direct signal. In conventional correlator simulations, code phase error and carrier phase error are computed with 1 chip early-late discriminator. The chip spacing of a narrow correlator is less than 1 chip. A spacing of 0.2 chips utilized to serve the discriminator functions. Two different narrow correlator discriminators are adopted in a strobe correlator and the chip spacing of the two narrow correlators can be adjusted to 0.1 and 0.2 chip. The same parameters are also adopted in both enhanced strobe and edge correlators. The E1/E2 tracker of the two correlators is located at E1=-0. and E2=-0. with 0.1 chip spacing (Irsigler, M. et al, 2003). The Adaline-based adaptive filter method with the parameter of tap delay $d\tau$= 0.01 chip, 0.1 chip, 0.5 chip and its 5 delayed tap is employed as the input to the filter. The number of multi-correlator is set as five. The initial learning rate is 0.05, the number of training samples is 5000 at 1ms C/A code period and the weights are initialized to 1. The performance is assessed on a separate test set of 100ms samples measured at intervals of 1ms samples during the adaptive process.

3.2. Performance Analysis and Comparison

With regard to crucial multipath mitigation techniques of internal receiver, the multipath performance of these correlation techniques will be compared with each other, including the proposed method. Thus, the envelopes of all techniques described above are plotted into the same diagram to make a comprehensive comparison of multipath mitigation performance

Figs. 10-12 compare the error envelopes of the code phase and carrier phase for all of the multipath mitigation techniques. Simulation results show that the proposed method with multipath delay at the location of set correlator as $d\tau$= 0.5 chip case has both the best multipath mitigation performance. Assume the location of correlator is not at multipath delay ($d\tau$ = 0.1 chip), it also presents good performance. The conventional PLL has a maximum 0.52 radians in carrier phase error. Therefore, the use of the conventional correlator can yield very large maximum multipath errors and reveals the worst mitigation performance. The same results hold true for both narrow and edge correlators. Note that since the narrow, the MEDLL, the edge and strobe correlators do not provide any carrier phase elimination, their sensitivity to multipath is almost the same as the 1-chip early-late correlator. Only slight differences can be observed due to differences in their code multipath mitigation.

These figures indicated that through the use of proposed method in combination with multi-correlator technique with a delay element dτ=0.5 chip, both code and carrier phase errors are reduced in the range of delay from 0 through 1.5 chip. In contrast, through the adoption of the proposed method with a tap delay dτ=0.5, the code and carrier phase error decrease significantly in the range of delay from 0 to 1.5 chip. The Figure shows that the use of multi-correlator technique can effectively reduce code phase error. Nevertheless, for carrier phase error, its performance remains the same. The reason is because carrier phase error is not related to multi-correlator. In the case of the tap delay dτ=0.5, multipath mitigation performance degrades in comparison with the case of dτ=0.1. This is because to the accuracy of the estimated delay profile in the Adaline-based filter depends on the tap delaydτdτ. The smaller the dτdτ the better the performance of multipath mitigation. In the case of the dτ=0.5 chip, the multipath mitigation performance degrades in code phase error and the carrier phase error also exceeds that of the conventional tracking loop. Though the use of a small tap delay can yield high performance in multipath mitigation, it also takes large computation cost to estimate delay profiles. However, the use of multi-correlator can save computation cost to estimate delay profiles.

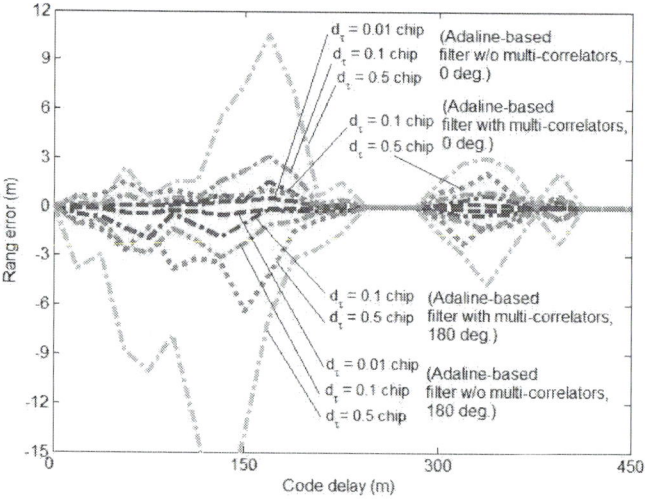

Figure 10. Code-phase error simulation results of proposed method. (α0=1.0α0=1.0,α1=0.5α1=0.5 ,τ0=0τ0=0 chip, τ1=0~1.5τ1=0~1.5chip, θ0=0∘θ0= 0∘, (θ1=0∘,180∘θ1=0∘,180∘) ; delay element dτ=dτ=0.01 chip, 0.1 chip, and 0.5 chip with and without multi-correlators.)

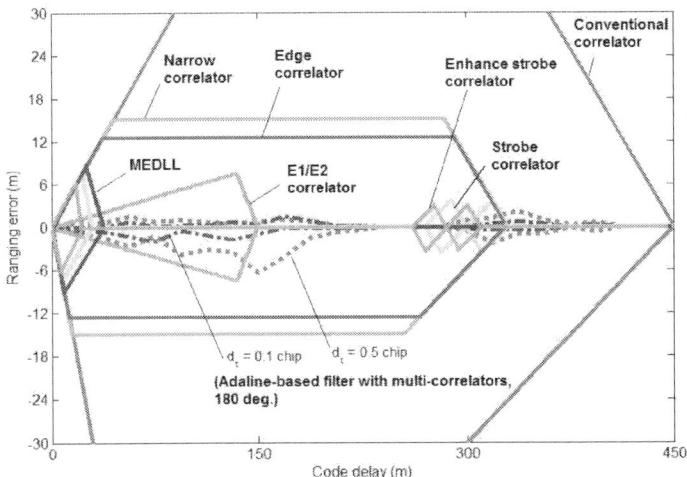

Figure 11. Code-phase error simulation results of existing methods. (α0=1.0α0=1.0,α1=0.5α1=0.5 ,τ0=0τ0=0 chip, τ1=0~1.5τ1=0~1.5chip, θ0=0∘θ0= 0∘, (θ1=0∘,180∘θ1=0∘,180∘)).

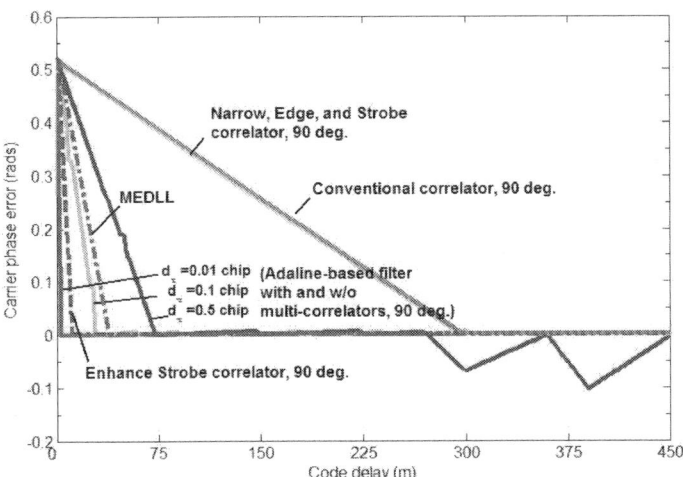

Figure 12. Carrier phase error simulation results. (α0=1.0α0=1.0,α1= 0.5α1=0.5 ,τ0=0τ0=0 chip, τ1=0~1.5τ1=0~1.5chip, θ0=0∘θ0=0∘, (θ1=0∘,180∘θ1= 0∘, 180∘)) ; delay element dτ=dτ=0.01 chip, 0.1 chip, and 0.5 chip with and without multi-correlators.)

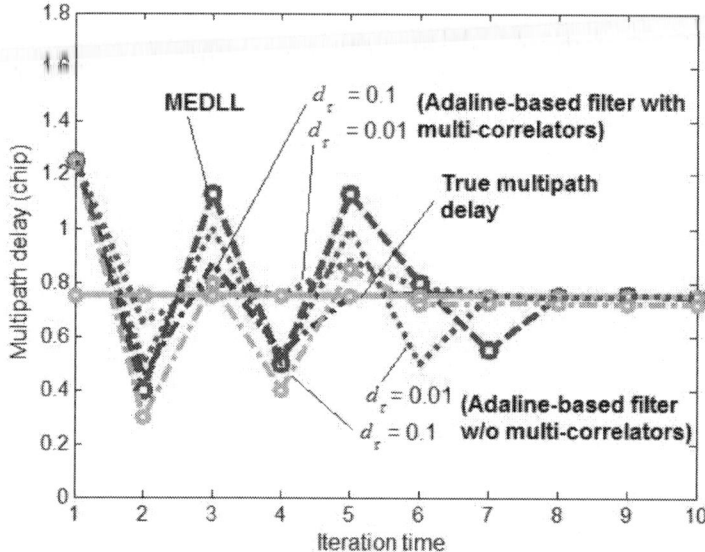

Figure 13. Delay estimated by MEDLL and proposed method with and without multi-correlators.

Note that Fig. 11 reveals that every DLL structure lacks of performance enhancement for short delay multipath signals. Nevertheless, the proposed method with multi-correlators can perform better in short delay and medium-to-long-delay multipath environment. Suppose a given application involves short delay and medium-to-long-delay multipath, then the best correlation techniques such as the enhanced strobe correlator will not outperform the proposed method of this paper.

To accomplish estimated performance of proposed method, the desired correct of multipath delay profiles are set $\alpha_0=1.0,\ \alpha_1=0.5,\ \tau_0=0,\ \tau_1=0.75,\ \theta_0=0°\,\text{and}\,\theta_1=0°$. The delay element number is five and the number of multi-correlator is set as five. An estimated multipath delay versus the true multipath delay curve for three considered algorithms, the MEDLL and the Adaline-based filter with and without multiple correlators, is shown in Fig. 13. Note that the proposed method with and without multi-correlator technique of dτ=0.01 has the faster convergence rate than the MEDLL. The Adaline-based filter without multi-correlator technique is rapid in convergence rate with dτ=0.1. However, it suffers from a steady state error 0.03 chip in delayed estimation. Nevertheless, the use of five multi-correlators with shift 0.1 chip, the error approximates zero.

Several concessions exist in these architectures such as: noise performance, code versus carrier performance, a priori information needed as input, short delay performance and hardware/software complexity. These factors are compared in Table 1. This performance comparison is on the basis of the published methods and simulation results of this paper. The research analysis is shown in the following:

Concerning the noise mitigation performance, when SNR =-10dB, the simulation result presents that the narrow correlator and proposed method with multi-correlators are the best in performance with the code tracking error of about 0.034 chip and 0.04 chip, respectively. The proposed method without multi-correlators in this paper is medium in performance with the tracking error of around 0.05~0.1 chip, which equals the medium noise performance of the edge and E1/E2 correlator. In contrast, the conventional correlator, strobe, enhanced strobe correlator and the MEDLL are inferior in noise performance, with the tracking error around 0.2 chip.

In term of the GPS mobile applications, high precision is required even at the expense of slightly increased complexity. The best options are the enhanced strobe correlator and the Adaline-based filter. The proposed method has the best performance in multipath mitigation. Nevertheless, its hardware complexity, such as the number of required multiplications per delay estimate is on the order of $O\left[N_{iter}(Kd_\tau)^3\right]$ where N_{iter} is the number of filter iterations and Kd_τ is an estimate of the maximum delay spread of the channel in the samples. The high complexity of this method is principally due to the matrix inversion operations. However, in short-delay and medium-to-long delay multipath environments, the number of delay samples Kd_τ are smaller. Thus, the complexity of the Adaline-based filter is not very high. The enhanced strobe correlator has lower complexity, on the order of $O\left[(Kd_\tau)^2\right]$, but it does not perform as well as Adaline-based filter. With regard to design perspective, the best tradeoff between accuracy and complexity should be determined based on estimated maximum delay spread of the channel.

Concerning the conventional receiver design for civilian application, the lowest complexity solutions of the 1-chip E-L correlator and the narrow correlator, appear to be the best choice. What is more, complexity is the top priority and is emphasized more than performance in the design of a receiver provided that no significant degradation in performance occurs. All of the conventional, strobe and narrow correlator designs have least medium performance and reduced complexity in multipath scenarios. Hence, they are viable options for a low complexity receiver. In comparison, even though the edge, the E1/E2 and the MEDLL designs are higher in cost, they are better than the conventional, narrow, and strobe correlators in performance.

In fact, there are inherent limitations in almost every technique. Note that the combined features of proposed method prevails over those of other techniques. In addition, the condition of short-delay and medium-to-long-delay multipath renders the effect of hardware complexity in Adaline-based filter insignificant. Consequently, the proposed method is a well-suited and well-balanced application in multipath mitigation.

TABLE 1. Comparative performance of multipath mitigation techniques. A: Best, O: Good, F: Fair, X: Poor

	Conventional correlator			Enhanced strobe				Adaline-based filter	
	Conventional correlator	Narrow	Strobe	Enhanced strobe	Edge	E1/E2	MEDLL	Without multi-correlators	With multi-correlators
Code multipath performance	X	F	F	O	F	O	O	A	A
Carrier multipath performance	X	X	X	F	X	X	O	F (Count on Number d_t)	A
Short-delay multipath performance	X	X	X	F	X	O	O	A	A
Medium-to-long-delay multipath performance	X	X	O	O	F	A	A	F	A
A priori information	Yes (Coarse delay)	Yes (Coarse delay)	Yes (Coarse delay)	Yes (Coarse delay)	Yes (Coarse delay)	Yes (Coarse delay)	Reference function	None	None
Noise	X	O	X	X	F	F	F	F	O
Performance (SNR= -10dB)	(above 0.2 chip error)	(0.034 chip error)	(0.2~0.25 chip error)	(below 0.2 chip error)	(0.054 chip error)	(0.04~0.06 chip error)	(below 0.18 chip error)	(0.05~0.1 chip error)	(0.03~0.05 chip error)
Hardware complexity	Easy	Easy	Med	Med	Med	Med	Hard	Hard (Count on number of iteration)	Hard
Software complexity	Easy	Easy	Easy	Hard	Med	Easy	Med	Med	Med

4. CONCLUSION

Multipath is the primary error source in high-precision-based GNSS applications and is also a significant error source in non-differential applications. Various receiver designs have been on the market and claim various multipath mitigation functions. Most of these techniques can be characterized either as discriminator function shaping or correlation function shaping. In this paper, an Adaline-based filter with multi-correlators method is adopted in multipath mitigation for GNSS application. A simplified direct plus multipath signal model is employed in this simulation. This approach enhances the performance of code phase and carrier phase errors compared with other published methods. Simulation results demonstrates that the proposed method is a viable and effective solution to increase the positioning accuracy for GNSS navigation in the presence of short-delay and medium-to-long-delay multipath environment.

5. ACKNOWLEDGEMENTS

The author would like to thank the National Science Council of Taiwan (R.O.C.) for their support of this work under grant NSC 99-2221-E-020-036.

REFERENCES

1. C. L. Chang, J. C. Juang, 2008 An Adaptive Multipath Mitigation Filter for GNSS Applications," EURASIP Journal on Advances in Signal Processing, 2008 Article ID 214815, 10 pages

2. L. Garin, F. van Diggelen, J. Rousseau, 1996 Strobe and Edge Correlator Multipath Rejection for Code, in Proc. ION GPS-96, 657 664 , Kansas City, MO, September 17-20

3. L. Garin, J. Rousseau, 1997 Enhanced Strobe Correlator Multipath Rejection for Code and Carrier, in Proc. ION GPS-97, 559 568 , Kansas City, MO, September 16-19

4. L. L. Hagerman, 1973 Effects of Multipath on Coherent and Non-coherent PRN Ranging Receiver, Aerospace Report TOR-0073 (3020-03)-3, Development Planning Division, The Aerospace Corporation

5. S. Haykin, 1986 Adaptive Filter Theory, 0-13048-434-2 Hall, USA.

6. M. Irsigler, B. Eissfeller, 2003 Comparison of multipath mitigation techniques with consideration of future signal structures, in Proc. ION-GPS/GNSS, 2584 2592 , Portland, OR, USA, September 9-12

7. R. A. Jacobs, 1988 Increased Rates of Convergence Through Learning Rate Adaptation, Neural Networks, 1 295 307

8. P. Mattos, 1996 Multipath Elimination for the Low-Cost Consumer GPS, in Proc. ION GPS-96, 665 671 , Kansas City, September 1996.

9. D. E. Rumelhart, G. E. Hinton, R. J. Williams, 1986 Learning Internal Representations by Error Propagation, Parallel Distributed Processing, 1 MIT Press, Cambridge, MA, USA, 318 362

10. R. J. Schalkoff, 1997 Artificial Neural Networks, 0-07115-554-6Hill.

11. J. Sleewaegen, M. , F. Boon, 2001 Mitigating Short-delay Multipath: A Promising New Technique, 204 213 , in Proc. ION GPS 2001, Salt Lake City, UT, USA

12. J. Stone, M. Chansarkar, 2004 Anti-multipath Triangulation (AMT) for Positioning in Dense Urban Environments, in Proc. ION GNSS, 1165 1168 , Long Beach, CA, September 21-24

13. J. van Dierendonck, P. Fenton, T. . Ford, Fall, 1992 Theory and Performance of Narrow Correlator Spacing in GPS Receiver, Navigation: Journal of the Institute of Navigation, 39 3 265 283

14. A. J. van Dierendonck, M. S. Braasch, 1997 Evaluation of GNSS Receiver Correlation Processing Techniques for Multipath and Noise Mitigation, in Proc. ION-NTM, 207 215 , Santa Monica, CA, USA, January 14-16

15. R. D. J. van Nee, 1992 The Multipath Estimating Delay Lock Loop, in Proc. IEEE Second Symposium on Spread Spectrum Techniques and Applications, 39 42

16. R. D. J. van Nee, J. Siereveld, P. C. Fenton, B. R. Townsend, 1994 The Multipath Estimating Delay Lock Loop: Approaching Theoretical Accuracy Limits, in Proc. IEEE PLANS, 246 251 , Las Vegas, Nev, USA, April 11-15

17. B. Widrow, Hoff, M. E. , 1960 Adaptive Switch Circuits. In IRE WESCON Convention Record, 96 104 , New York, USA

18. B. Widrow, M. A. Lehr, 1990 30 Years of Adaptive Neural Networks: Perceptron, Madaline, and BP., Proc. IEEE, 1550 1560

Chapter 10

Adaptive Data Filtering of Inertial Sensors with Variable Bandwidth

Mushfiqul Alam [*] and Jan Rohac

Department of Measurement, Faculty of Electrical Engineering, Czech Technical University in Prague, Technicka 2, Prague 16627, Czech Republic

ABSTRACT

MEMS (micro-electro-mechanical system)-based inertial sensors, *i.e.*, accelerometers and angular rate sensors, are commonly used as a cost-effective solution for the purposes of navigation in a broad spectrum of terrestrial and aerospace applications. These tri-axial inertial sensors form an inertial measurement unit (IMU), which is a core unit of navigation systems. Even if MEMS sensors have an advantage in their size, cost, weight and power consumption, they suffer from bias instability, noisy output and insufficient resolution. Furthermore, the sensor's behavior can be significantly affected by strong vibration when it operates in harsh environments. All of these constitute conditions require treatment through data processing. As long as the navigation solution is primarily based on using only inertial data, this paper proposes a novel concept in adaptive data pre-processing by using a variable bandwidth filtering. This approach utilizes sinusoidal estimation to continuously adapt the filtering bandwidth of the accelerometer's data in order to reduce the effects of vibration and sensor noise before attitude estimation is processed. Low frequency vibration generally limits the conditions under which the accelerometers can be used to aid the attitude estimation process, which is primarily based on angular rate data and, thus, decreases its accuracy. In contrast, the proposed pre-processing technique enables using accelerometers as an aiding source by effective data smoothing, even when they are affected by low frequency vibration. Verification of the proposed concept is performed on simulation and real-flight data obtained on an ultra-light aircraft. The results of both types of experiments confirm the suitability of the concept for inertial data pre-processing.

Keywords: inertial navigation; attitude control; filtering algorithms; adaptive signal processing; accelerometers

1. INTRODUCTION

Recently, there has been a growing trend toward using cost-effective MEMS (micro-electro-mechanical system) technology-based sensors for navigation purposes in aerospace systems, such as on light aircrafts and unmanned aerial vehicles (UAVs). A strapdown inertial system consisting of tri-axial accelerometers (ACCs) and tri-axial angular rate sensors (ARSs) is commonly used for attitude estimation (roll, pitch, yaw angle), as well as for velocity and position evaluations. On large aircraft, ring laser gyros and servo ACCs are used on board for precise measurements, which is expensive. In comparison, MEMS sensors are compact, lightweight and cost effective, thus offering an inexpensive solution for navigation purposes. However, at the same time, MEMS-based inertial sensors suffer from bias instability, insufficient sensitivity, noise, *etc.*, which present significant challenges in data processing that have to be dealt with in navigation processes. Originally, the attitude was supposed to be evaluated by integrating angular rates; nevertheless, as mentioned before, the measurements suffer from several inaccuracy impacts. In the case of the ARS-based attitude evaluation process, this inaccuracy causes unbound error growth, which needs to be corrected by data obtained from so-called aiding systems, e.g., magnetometers, cameras and even ACCs. These aiding systems provide information about attitude only under certain conditions, limiting their usability. This paper focuses on data pre-processing for navigation solutions based on inertial sensors only (ARSs and ACCs). Therefore, the ARS-based attitude evaluation process is primarily aided by ACC-based attitude evaluation [1]. This aiding can be applied under conditions when only gravity affects ACC measurements and no other acceleration is present [2–4]. This principle is common in cost-effective solutions of attitude and heading reference systems (AHRSs); however, these ideal aiding conditions are hardly achievable in harsh environments, due to strong vibrations present on light or small aircrafts. This complicates the situation, as the frequency of those vibrations cannot be simply isolated from the aircraft dynamics. The ARSs are primarily used for attitude evaluation, unlike the ACCs, which are utilized in AHRS just for attitude compensation.

To learn the characteristics of real flight conditions, several flight experiments were performed using IMU ADIS16350 (Analog Devices, Norwood, MA, USA), which was utilized in the EFIS INTEGRA TL-6524 (Electronic Flight Instrumentation System) ,flight monitoring system manufactured by TL-Elektronic, Inc. (Hradec Králové, Czech Republic) The system was mounted to the instrument panel of the ATEC321 aircraft (ATEC321 is a Czech ultra-light aircraft, designed and produced by ATEC v.o.s, Libice nad Cidlinou, Czech Republic). The instrument panel was equipped neither with active nor passive vibration dampers. As a result, the sensors were directly affected by strong structural vibrations. Measurements were made for different flight phases, such as parking, taxing on the runway, taking off, during the flight and landing. The data were recorded at a sampling frequency of 43 Hz. The worst situation corresponds to the case when the vibration impact cannot be distinguished and isolated from the aircraft dynamics. Such a situation is depicted in Figure s 1 and 2, which show the flight data from ACC and ARS measured during the engine revolution per minute (RPM) suppression and their frequency spectrum.

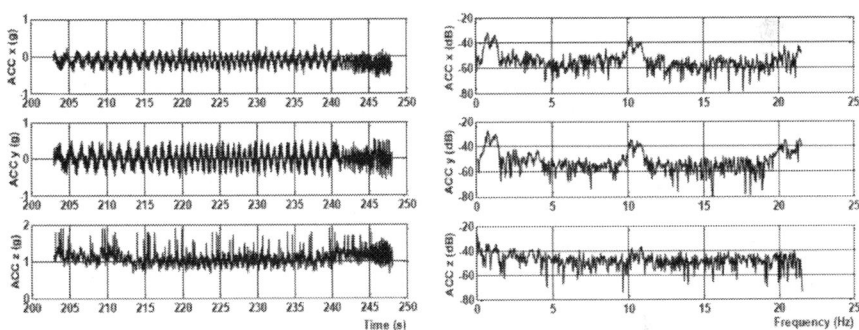

Figure 1. Accelerometer (ACC) measured during engine suppression and the frequency spectrum.

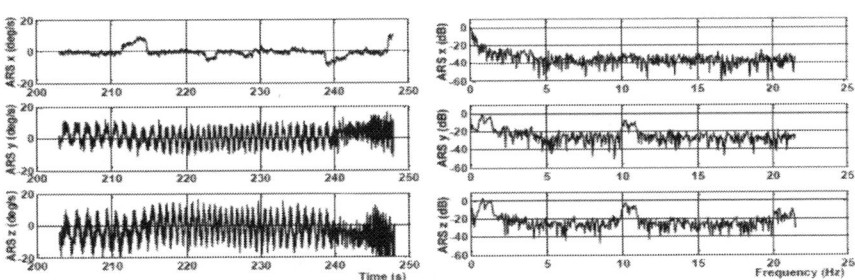

Figure 2. Angular rate sensor (ARS) measured during engine suppression and the frequency spectrum.

In cases of engine RPM suppression during the flight, vibration frequencies go down all the way to 0.5 Hz. ACCs are generally affected by the combination of translation, centrifugal and gravitational accelerations along with the vibrations arising from the propeller and the aircraft structure. The vibration effect often dominates the ACC measurements. In contrast, vibrations have slight impacts on ARS readings depending on g and g^2 sensitive parameters; contrariwise, their readings are affected by bias instability and noise. These different characteristics of ARS and ACC enable their data fusion to improve the final accuracy of the whole attitude estimation process. Generally, ARS data are always used to estimate attitude, even under dynamic conditions, when the aircraft is maneuvering or under steady flight conditions. Unlike ACC, data are directly utilized for the attitude compensation under only steady-state conditions when the gravity distribution in the sensor's framework can be estimated. This corresponds to situations in which the aircraft performs a direct and unaccelerated flight. In the cases where the aircraft undergoes a banked turn or a circular flight, a long-term additional acceleration is present due to the centripetal force created by traveling along a curved path. In such conditions, it is possible to estimate the centripetal acceleration and to subtract it out by providing the known velocity or airspeed to the data fusion process [5]. There exist several approaches to data fusion for attitude estimation, such as temporally-interconnected observers (TIO) [6], complementary filters [7] or Kalman filters [8,9]. However, the accuracy of the estimation is always reduced, while the ACC data are affected by periodic vibration. This complicates the situation, due to a harsh environment causing structural vibrations, which are directly picked up by the ACCs. Therefore, for precise attitude estimation regardless of the aircraft flight condition, it is essential to provide acceleration data that are as smooth as possible with a reduced vibration effect; thus, ACC data require pre-processing. The light aircrafts are classified as Level I Class I, Category B Flight Phase (cruise, climb, descent, loiter) by the Federal Aviation Administration (FAA). The maximum time to achieve a change in the bank and pitch angle is 1.7 s, and the minimum time is 0.2 s [10,11]. This means that the aircraft's operational frequency lies in the range of 0.6 up to 5 Hz. In this instance, the ideal choice would be to apply a band-pass (BP) filter; nevertheless, such a narrow bandwidth would require a very high order filter, which is not desirable for navigation purposes, because of the long delays.

As mentioned, the aircraft dynamics lies in the range of 0.6 up to 5 Hz, and the vibration frequency might go all the way down to 0.5 Hz. Therefore, using a constant 5 Hz bandwidth low-pass (LP) filter would mean that the 0.5 Hz frequency vibrations will not be filtered. On the other hand, using a constant 0.6-Hz bandwidth LP filter would lead to a situation in which the aircraft's motion information in the bandwidth up to 5 Hz would be lost. Therefore, our contribution is a novel concept of pre-processing ACC data using adaptive bandwidth filtering, which is modified based on sinusoidal data estimation. The proposed filtering algorithm is adaptive in the sense that the filtering bandwidth is modified based on the signal history. This enables the usage of ACC-based attitude compensation, even under variable low-frequency vibration impacts, while common commercial AHRS systems fail to have the correct compensation capability. This proposed approach brings several advantages against the ones commonly used, such as a smaller and acceptable delay, even when the narrowest bandwidth of 0.5 Hz is applied on the ACC data. On the other hand, ARS data are filtered with a constant bandwidth, and thus, when no low-frequency vibrations arise, all data are filtered with the same bandwidth of 5 Hz, which provides the same delay of data pre-processing for the majority of the flight. This approach hence brings an added advantage to inertial data pre-processing and enhances the ACC-based attitude compensation possibilities.

The rest of the paper is organized as follows: Section 2 outlines the methodology of the proposed concept in detail. A detailed description of the principle of estimating the sensor's signal via a sinusoidal estimation filtering algorithm is also presented in this section. Section 3 provides the results of the proposed algorithm applied on simulated data and real flight data and confirms the suitability of the approach. Section 4 concludes the paper with final remarks.

2. METHODOLOGY

This paper proposes an adaptive variable bandwidth filtering via sinusoidal data estimation to pre-process the data of ACCs and ARSs by as narrow a bandwidth LP filter as possible, while preserving the dynamics information included in the data. In the past, several attempts were made to use variable bandwidth filtering in communications systems [12,13], but the use was limited to a fixed length of finite impulse response (FIR) filters. However, the length of the filter is always proportional to its delay, which restricts the usability for navigation purposes. In the proposed concept, two key assumptions are made:

1. The vibration content in the inertial data has an approximately periodic and sinusoidal characteristic.

2. The frequency of the signal content varies gradually, and the changes are smooth.

A schematic block diagram of the proposed signal filtering method using variable bandwidth filters is depicted in Figure 3.

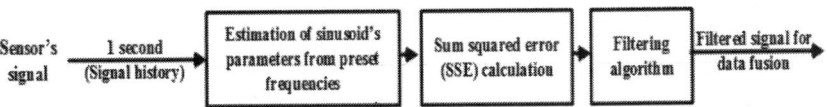

Figure 3. Complete filtering block diagram.

In general, the overall filtering process can be broken down into three main stages: (1) estimation of the sinusoid's parameter; (2) sum squared error calculation; and (3) applying filtering on the signal. The overall filtering task is carried out with a 1 s window of the signal history. This particular length of the history is chosen to have the capability to detect vibration content down to 0.5 Hz, since at least a half cycle of the approximate sinusoidal signal is necessary for the estimation process. The estimation of the sinusoid's frequency is based on preset frequencies, which are chosen according to the required filtering pass bands and bandwidths associated with the flight operational conditions. When raw signals enter the first block, they are fitted with sinusoids of all preset frequencies to get their approximations. The raw signals and their sinusoidal approximations are then led to the second block to calculate the sum squared error (SSE) of the fitting. The sinusoidal approximation with the preset frequency for which the SSE is the lowest marks the best fit and, thus, indicates the operating frequency of the vibration's strongest content. Finally, based on the operating frequency, a variable bandwidth in the filtering algorithm is adapted and applied on the signal. The process in the first and second block can be easily performed by fast Fourier transformation (FFT) while post-processing; however, it is computationally expensive for real-time applications. Thus, applying the proposed sinusoidal approximation technique brings an advantage in terms of lower computational demands, making the filtering suitable for real-time applications. Details about the chain of signal processing described above are presented in the following subsections.

2.1. Principles of Sinusoidal Signal Estimation

As mentioned earlier in Section 1, the vibration impact on the sensor's signal is often periodic, sinusoidal in nature and with one strongest frequency content. Therefore, it is often a reasonable approximation to address the problem of determining/estimating the frequency content in the raw signal via sinusoidal fitting algorithms.

Assume that the signal history x is obtained at time instances t where N is the total number of samples in the sequence. N is chosen to preserve a 1 s window. x_N and t_N correspond to the latest sample, and $(N - 1)$ down to 1 represent samples in the signal history.

$$x_n = [x_1 \, x_2 \, ... \, x_{N-1} \, x_N]^T; \quad t_n = [t_1 \, t_2 \, ... \, t_{N-1} \, t_N] \qquad (1)$$

For a small signal history window, the history sequence x can be assumed as periodic with a frequency f and angular frequency $\omega = 2\pi f$. The orthogonality relationships of the sine and cosine functions can be used to break down an arbitrary periodic function into a set of simple terms that can be summed, solved individually and then recombined to obtain the solution to the original signal sequence x_n or to its approximation. Using the method for a generalized Fourier series, the signal history sequence x can be represented or modeled as a summation of the cosine and sine, as follows:

$$s_n(\vartheta) = A\cos(\omega \times t_n) + B\sin(\omega \times t_n) + C \qquad (2)$$

where s_n is the approximation of x_n; A, B, C, ω and f are unknown constants, and ϑ defines the set of these four unknown parameters (A, B, C and ω). The sine wave fitting problem in (1) and (2) can be then solved by minimizing of the sum squared error [14,15], which is given by:

$$V(\vartheta) = \frac{1}{N}\sum_{n=1}^{N}(x_n - s_n(\vartheta))^2 \qquad (3)$$

Consider the particular parameter vector ϑ; where $\vartheta = [A\ B\ C]^T$ and ϑ can be written as:

$$\vartheta = [\theta^T\ \omega]^T$$

Let $D(\omega)$ be the $N \times 3$ matrix defined as:

$$D(\omega) = \begin{bmatrix} \cos(\omega \times t_1) & \sin(\omega \times t_1) & 1 \\ \vdots & \vdots & \vdots \\ \cos(\omega \times t_N) & \sin(\omega \times t_N) & 1 \end{bmatrix} \qquad (4)$$

The sum squared error in (4) can be written as:

$$V(\vartheta) = V(\omega, \theta) = \frac{1}{N}\{[x_n - D(\omega)\theta]^T[(x_n - D(\omega)\theta)]\} \qquad (5)$$

When the frequency f of the signal history is known (in other words, angular frequency ω is known), Equation (5) can be minimized in the least squares sense by solving the set of linear equations $D(\omega)\vartheta = x_n$ [16]. If $D(\omega)$ has the full rank, the solution of the estimated $\hat{\vartheta}$ is given by:

$$\hat{\theta} = \begin{pmatrix} \hat{A} \\ \hat{B} \\ \hat{C} \end{pmatrix} = \left(D(\omega)^T D(\omega) \right)^{-1} D(\omega)^T x_n \tag{6}$$

It can be noted that for large N, the columns in $D(\omega)$ become orthogonal. Thus, $D(\omega)^T D(\omega)$ becomes a diagonal matrix with elements $[^N/_2 \ ^N/_2 \ N]$. Thus, it makes the calculation of the inverse of $(D(\omega)^T D(\omega))$ to estimate $\hat{\theta}$ computationally inexpensive.

2.2. Sum Squared Error Calculation to Estimate the Filtering Bandwidth

The principles described in Section 2.1 are used to estimate the frequency content in the signal considering the 1 s window of the signal history. The mentioned Equations (5) and (6) can be solved easily when the frequency f is known. Therefore, the proposed approach uses preset frequencies $\hat{f}_i$. These frequencies specify the filtering bandwidth, which can be then applied in the third block in Figure 3. Each $\hat{f}_i$ is used to estimate $D(\omega)$ defined in Equation (4) and $\hat{\theta}$ in Equation (6). The sinusoidal signal $\hat{x}_u$ is then constructed for each $\hat{f}_i$ and the corresponding $\hat{\theta}$ using:

$$\widehat{x_{n_i}} = \hat{A} \cos(2\pi \hat{f}_i \times t_n) + \hat{B} \sin(2\pi \hat{f}_i \times t_n) + \hat{C} \tag{7}$$

This step gives the advantage of using as many estimates as needed for the specific application and required filtering bandwidths. To get the best sinusoidal approximation with respect to the original raw signal x_n, the SSE is calculated using:

$$SSE = \sum_{1}^{N} \left(x_n - \widehat{x_{n_i}} \right)^2 \tag{8}$$

The preset frequency $\hat{f}_i$, which gives the lowest value of the SSE, provides the best fit for the measured signal and, thus, indicates the operating frequency based on which the filtering bandwidth of the third stage is adapted and applied. Since the estimation process uses the 1 s window, the proposed approach has a corresponding learning time of 1 s. This means that the proposed approach will take 1 s to detect a complete change/transition in the frequency and to adapt the filtering bandwidth.

2.3. Filtering Algorithm

A conventional finite impulse response (FIR) filtering (such as *generalized equiripple*, quadratically weighted moving average, *etc.*) alone cannot be used for such a low filtering bandwidth (≈0.5 Hz) while providing smooth data. In addition, reaching such a narrow bandwidth would lead to higher order filters, which is not desirable, since they produce a long delay in the signal processing. Therefore, a novel multistage adaptive filtering approach is developed, as demonstrated in the block scheme in Figure 4. The proposed filtering process is adaptive in the sense that the bandwidth of the overall filtering process can vary with respect to the frequency content in the signal. The filtering process can be broken down into two main stages. The first stage is the filtering of the signal using a variable bandwidth Kaiser windowed filter with coefficients $[b_1\ b_2...\ \cdot\cdot\ b_{N-1}\ b_N]$, while the second stage utilizes an LP wavelet filter with a variable level of decomposition.

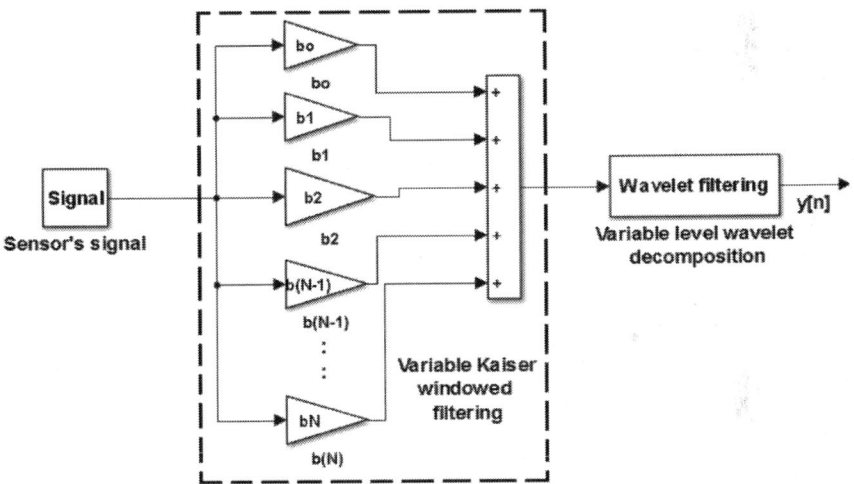

Figure 4. Schematic diagram of the new filtering algorithm.

For the first stage, a Kaiser windowed LP filter is chosen, since it allows to control the transition band, pass band and stop band ripples through a proper choice of the filter order and has further a unique *sin* function shape, which provides low bandwidth and low side lobes at an equivalent filter length compared to other types of conventional filters. The coefficients of the Kaiser windowed LP filter can be calculated as:

$$b(n) = \frac{I_o(\frac{2\beta}{M}\left(\sqrt{n(M-1)}\right)}{I_o(\beta)} \tag{9}$$

where β is an arbitrary, non-negative real number that determines the shape of the window. I_0 is the zeroth order modified Bessel function and N is the order of the filter.

The second stage is formed by a wavelet filter, which brings further advantages in terms of a further attenuation of high frequency noise, while preserving an acceptable delay. Different levels of wavelet decomposition can be reached by taking an LP mother wavelet filter, upsampled by a factor of 2 and convolving it with the same LP mother wavelet filter. This process can be repeated to achieve different levels of decomposition and, thus, different filtering bandwidths. In our case, the *sym*4 mother LP wavelet was considered. The details of the choice of mother wavelet filter and for obtaining different levels of decomposition are outlined in [17–20].

The filtering bandwidth of wavelet filtering cannot be explicitly chosen or controlled; however, wavelet filters are capable of providing smooth data for signal reconstruction. Whereas the filtering bandwidth for the Kaiser window can be chosen based on Equation (9), the two characteristics of the two filters can be combined together to provide one overall filter that is efficient in low frequency attenuation, while keeping the filter order minimal; in other words, keeping the delay minimal.

In the filtering process, the raw signal is passed through the Kaiser windowed LP filter and then filtered by the wavelet filter to further suppress the high frequency noise and to smooth the filtered signal. Note that the overall filtering bandwidth can be modified by varying the length of the Kaiser windowed LP filter and by modifying the level of the wavelet filter decomposition. Simulation results are discussed in detail in Section 3.2, and the experimental verification is in Section 3.4.

3. PERFORMANCE EVALUATION AND DISCUSSION

The performance of the proposed filtering approach is evaluated based on simulated data, as well as on real flight data. The algorithm was implemented using MATLAB. The results are presented in detail in the following subsections.

3.1. Performance of the Filtering Algorithm with Different Bandwidths

As mentioned above, a main objective of the proposed filtering algorithm is to achieve efficient filtering performance in terms of low frequency vibration attenuation in the signal while preserving an acceptable delay. For this reason, we have split the frequency range of interest, 0.5 to 5.5 Hz, into 11 bandwidths with a step size of 0.5 Hz. Based on the required bandwidth values, we have optimized the coefficients of the Kaiser windowed LP filter using Equation (9) and chose the levels of wavelet filter decomposition.

To observe the efficiency of the overall filtering, we simulated a sinusoidal signal with frequencies in the range of 0.5 to 14 Hz and let it pass through both stages of the proposed filtering algorithm. The sampling frequency of the simulated signals was chosen to be 43 Hz to be consistent with the sampling frequency of the real flight experiment. The performance of the filtering for two signal frequencies (0.5 Hz and 1 Hz) when the bandwidth was set to 0.5 Hz is depicted in Figure 5. For all combinations of the bandwidths and the signal frequencies, see Table 1, which summarizes the filtering performance. Particular delays corresponding to the filtering performances are denoted in Table 2. It can be seen that the maximum delay in the signal is about 0.39 s, which is an acceptable delay for ACC signal processing and attitude compensation.

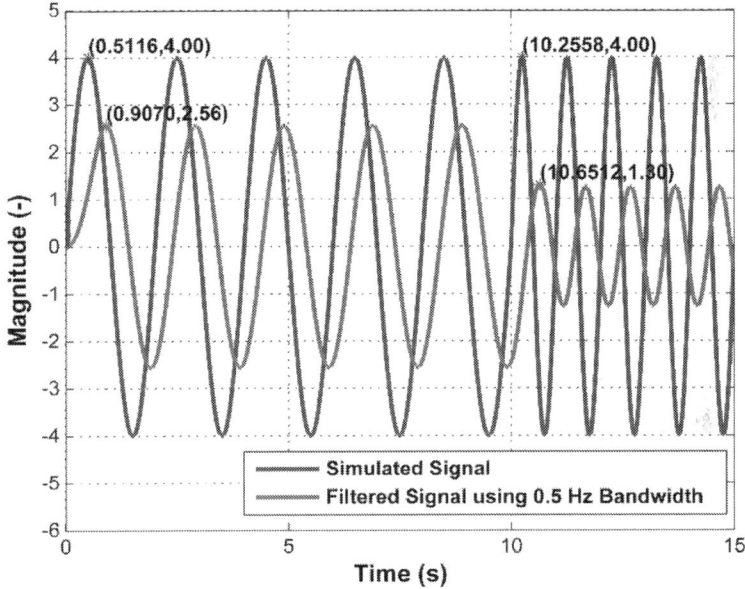

Figure 5. Filtering the simulated signal using only the 0.5 Hz bandwidth filter.

Table 1. Filtering efficiency at different filtering bandwidths for various signal frequencies.

	Signal Frequency (Hz)															
	14	12	10	8.0	6.0	5.0	4.5	4.0	3.5	3.0	2.5	2.0	1.5	1.0	0.6	0.5
5.5 (M = 24, L = 1)	−34.5	−31.5	−23.3	−9.0	−6.0	−0.5	0.1	0.3	0.0	−0.1	−0.3	−0.2	−0.1	−0.1	0.1	0.1
5.0 (M = 28, L = 1)	−45.7	−35.2	−29.9	−24.1	−8.9	−3.7	−0.5	0.5	0.4	0.0	−0.1	−0.2	0.0	−0.1	0.0	0.0
4.5 (M = 30, L = 1)	−42.7	−38.3	−32.6	−31.0	−25.7	−6.2	−2.5	−0.6	0.0	0.1	0.0	−0.4	−0.4	−0.3	0.0	0.0
4.0 (M = 26, L2)	−42.6	−38.6	−33.7	−28.8	−24.1	−10.1	−7.4	−3.4	−0.6	0.0	0.1	0.0	−0.3	−0.4	−0.1	0.0
3.5 (M = 29, L2)	−36.7	−35.0	−32.8	−30.1	−23.4	−22.5	−15.4	−7.7	−3.7	−0.9	0.5	0.5	0.5	0.5	0.1	0.1
3.0 (M = 32, L = 1)	−53.7	−40.8	−40.8	−36.5	−29.1	−26.1	−24.9	−26.6	−9.4	−3.7	−0.3	0.5	0.3	0.1	−0.1	0.0
2.5 (M = 28, L = 2)	−45.6	−44.7	−42.3	−37.5	−29.5	−31.1	−22.8	−23.0	−21.2	−12.8	−6.0	−1.2	−0.1	0.7	0.4	0.0
2.0 (M = 32, L2)	−41.5	−38.5	−43.0	−33.8	−37.3	−29.2	−36.3	−25.9	−28.0	−19.5	−9.4	−3.6	−0.6	0.0	0.1	0.1
1.5 (M = 7, L = 3)	−40.7	−40.1	−45.6	−29.9	−41.4	−37.5	−35.5	−34.5	−21.4	−12.7	−18.4	−4.2	−2.5	−0.9	0.0	0.0
1.0 (M = 38, L = 1)	−62.7	−62.5	−61.6	−59.4	−57.5	−56.4	−58.2	−56.4	−59.4	−43.3	−26.2	−15.3	−8.2	−3.6	−0.5	0.0
0.5 (M = 33, L = 3)	−39.2	−40.1	−35.8	−38.5	−40.0	−40.1	−29.4	−37.8	−27.3	−37.6	−24.1	−35.3	−20.5	−9.8	−4.5	−3.9

Note: M corresponds to the Kaiser windowed filter order, and L corresponds to the level of wavelet decomposition (LoD).

Table 2. Time delays for the chosen filtering bandwidths.

Filtering Bandwidth (Hz)	Time Delay (s)
5.5	0.23
5.0	0.24
4.5	0.26
4.0	0.33
3.5	0.35
3.0	0.36
2.5	0.37
2.0	0.38
1.5	0.38
1.0	0.39
0.5	0.39

The attenuation level in the filtered signal is calculated using:

$$G_{dB} = 20 \log_{10}(A_2/A_1) \tag{10}$$

where A_1 is the amplitude of the original signal and A_2 is the amplitude of the filtered signal.

The selection of the filtering bandwidth that is applied in the third block in Figure 4 is dependent on Tables 1 and 2. InTable 1, the minimum required level of attenuation corresponds to −15 dB; nevertheless, if further attenuation is needed, it is always a possibility to use a narrower bandwidth up to 0.5 Hz. Attenuation of −15 dB corresponds to approximately 1/5th of the original amplitude. The particular choice of filtering bandwidth based on the signal operating frequency is highlighted in dark grey in Table 1. For example, if the operating signal frequency is 10 Hz or higher, the filtering bandwidth is 5.5 Hz. If it is from 8 to 10 Hz, the filtering bandwidth is going to be 5 Hz.

3.2. Demonstration of Data Smoothing

As mentioned in Section 2.3, the wavelet filters are used to smooth the signal, as well as to further attenuate its unwanted high frequency content. To confirm these characteristics, a white noise was generated with variance set to unity. Two LP Kaiser filters were designed for the bandwidth of 2 Hz with different filter orders (M = 25 and M = 32) using Equation (10)and applied to the white noise. Afterwards, the signal filtered with the Kaiser filter of M = 25 was further passed through the wavelet filter with level of decomposition (LoD) = 1. The filtering results are shown in Figure 6. The results are further demonstrated in the frequency domain in Figure 7.

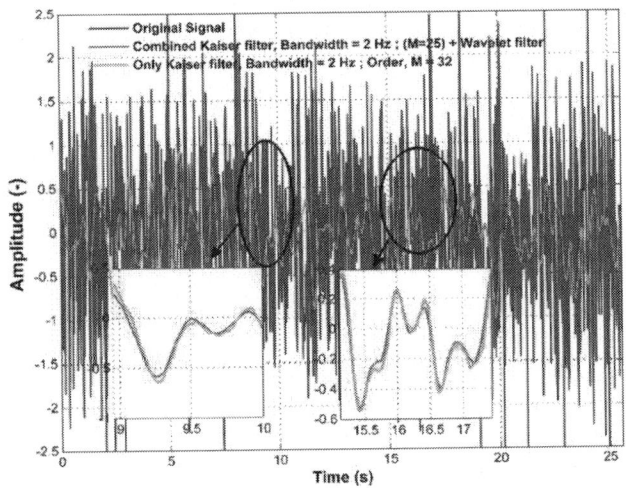

Figure 6. Comparison between filtering using only the Kaiser windowed low-pass (LP) filter and the proposed filtering multistage algorithm with the wavelet filter implemented in the time domain.

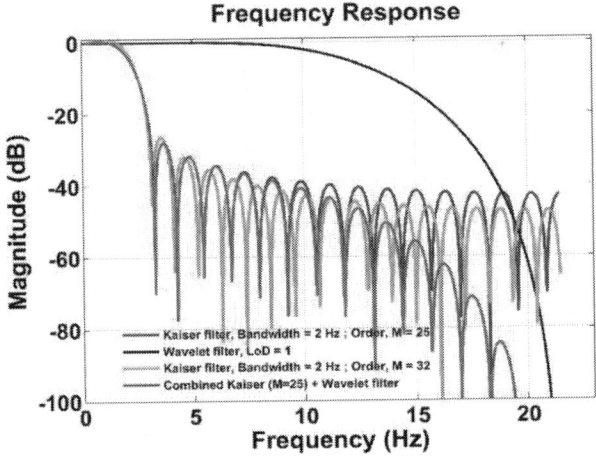

Figure 7. Comparison between filtering using only the Kaiser windowed LP filter and the proposed filtering multistage algorithm with the wavelet filter implemented in the frequency domain.

In Figure 7, it can be seen that both LP Kaiser filters with a cut-off frequency of 2 Hz have slightly different attenuation at higher frequencies; however, their delays are different, *i.e.*, 0.28 s and 0.36 s. Nevertheless, when the LP Kaiser filter of M = 25 (blue line) is combined with the wavelet filter with the first level of decomposition (LoD = 1) (black line), which has M = 7, the combined filter performance (red line) changes for the higher frequencies, while preserving the attenuation at low frequencies. The overall filtering has the order M = 32, and its filtering efficiency increases when attenuating higher frequencies. Thus, it reaches better performance than using only the LP Kaiser filter of M = 32 with a comparable time delay. This means that the proposed filtering approach provides a more enhanced filtering capability compared to conventional filters, while keeping the filter order at a minimum.

This can be explained by the wavelet filters being advantageous despite having an irregular shape to their frequency characteristics. They are able to perfectly reconstruct functions with linear and higher order polynomial shapes, such as rectangular, triangular, second order polynomials and windowed filters [21]. Note that Fourier series fail to do so while designing regular filters, such as Kaiser, and various other conventional filters mentioned earlier [22]. As a result, wavelets are able to denoise the particular signals better than conventional filters that are based on the Fourier transform design and that do not follow the algebraic rules obeyed by the wavelets.

3.3. Application of the Multistage Filtering Algorithm on Simulated Data

For testing the adaptability of the proposed filtering approach, a sinusoidal signal was simulated with two frequencies (2 Hz and 3 Hz) with additive white Gaussian noise of unity variance and a sampling frequency of 43 Hz. It was observed how the filter behaves with respect to a change of frequency. The resulting performance is shown in Figure 8.

The first part of Figure 8 contains the signal with 2 Hz, while 3 Hz is used in the second part. From the filtered data, it can be seen that the filtering method has a 1 s learning time to observe the complete transition between one operating frequency to another. The Figure also illustrates that for a low frequency (2 Hz), a lower filtering bandwidth is used than in the case of a frequency of 3 Hz. The zoomed part of the time range of 4.4 to 5.4 s shows the filter behavior when the filtering bandwidth is changed.

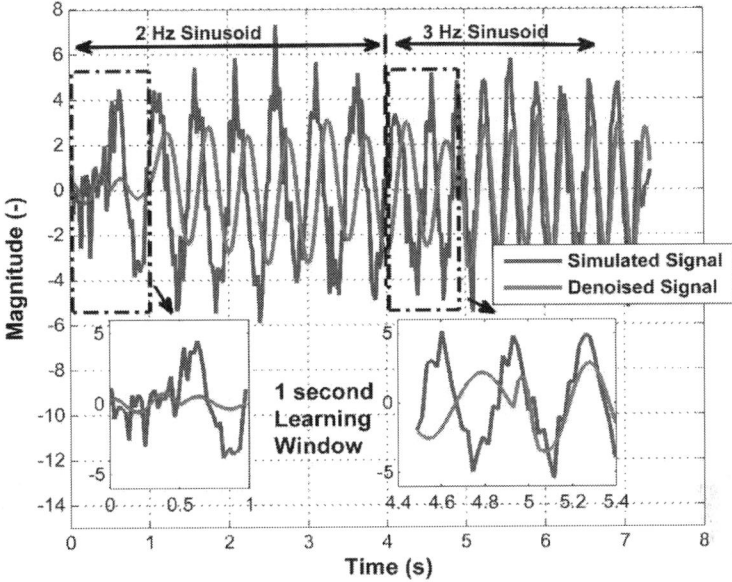

Figure 8. Filtering result on the simulated signal.

3.4. Application of the Algorithm on Real Flight Data

As mentioned in Section 1, a narrow bandwidth filtering is required due to the potential presence of low frequency vibrations affecting inertial sensors. Aircraft can fly under direct and un-accelerated conditions or under dynamic motion conditions. For cost-effective attitude estimation systems, it is necessary to consider that the signals from the ARSs require preserving the dynamics information, and in contrast, the ACCs' signals are used as an aiding source to compensate for the attitude estimation, only under steady flight conditions. Mentioned in Section 1, the motion dynamics of light aircraft lies within the 5 Hz bandwidth; hence, a constant bandwidth of 5.5 Hz is used for the ARSs' signal filtering. This choice provides unmodified dynamics in the range of 5 Hz, as required. In the case of ACC signals, an adaptive bandwidth filtering is applied to reduce the vibration effects in the signals. In other words, the same bandwidth of 5.5 Hz is chosen for both ARSs and ACCs when the flight conditions lead to noise and vibrations with frequencies higher than 10 Hz. This ensures the same delay for both ARSs and ACCs for the majority of the flight, which is advantageous. When conditions change and the vibration frequency goes down, the ACCs' signal filtering bandwidth is adapted accordingly, potentially down to 0.5 Hz, while the filtering bandwidth for the ARSs' signals does not change. This approach provides both observable dynamics from ARSs and a compensation capability for attitude estimation with the help of the

ACCs, even under steady flight conditions when the ACCs are affected by low frequency vibrations. Generally, due to the low frequency vibrations' influence, the operation of the majority of the commercially used cost-effective AHRSs is limited. In contrast, the proposed filtering approach is more robust while operating under a low frequency vibrating environment compared to the commonly used approaches.

Real flight data for ACCs and ARSs are obtained from flight experiments using the ultra-light aircraft, ATEC 321. The flight data were sampled at a frequency of 43 Hz. As proposed, for the filtering purposes, 11 different bandwidths were chosen at equal intervals from 0.5 Hz to 5.5 Hz. The filtering of the ARS data using a constant 5.5-Hz bandwidth LP filter is shown in Figure 9. It can be seen that the potential high frequency noise is attenuated and filtered in the ARS measurements, while preserving the delay up to 0.23 s.

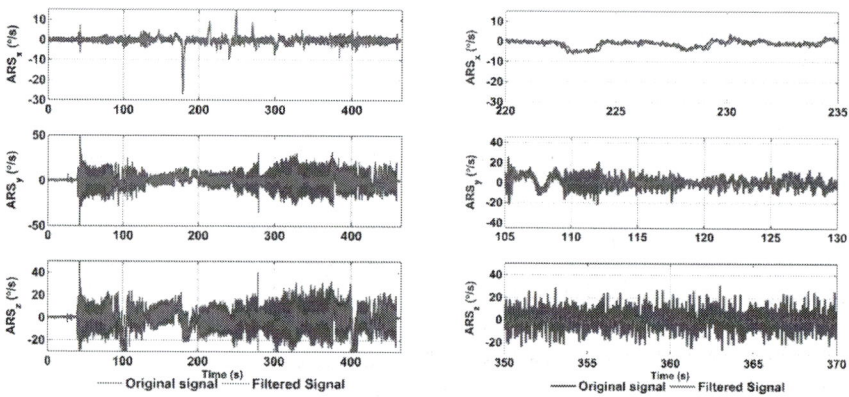

Figure 9. Angular rates measured during the flight, the filtered signals and the zoomed tracks.

Figure 10 shows the ACC signal from the same flight as shown in Figure 9.

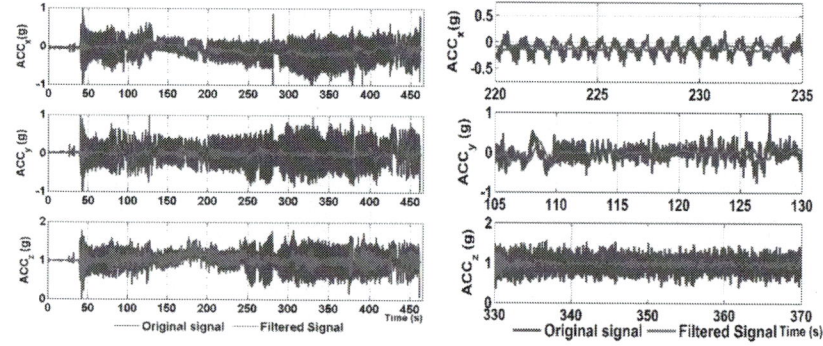

Figure 10. Acceleration measured during engine suppression, the filtered signal and the zoomed tracks.

It can be seen in Figure 10 that the low frequency vibrations affecting the signal during the engine suppression between 220–235 s are significantly attenuated in the case of ACC_x (zoomed tracks). This measured signal suffered from a vibration frequency of about 0.5 Hz, and thus, at this point, the filtering was performed at the lowest bandwidth corresponding to the highest filtering order. On the other hand, the ACC_y track in the range of 105–130 s suffered from high frequency vibrations; hence, the filtering was performed with a wider bandwidth; so, the filtering order was lower and the delay was shorter.

To confirm the adaptability of the filtering bandwidth, the ACC_x signal is shown in Figure 11 with zoomed parts. It demonstrates the variable bandwidth filtering capability and corresponding delays. In the left inset, the signal frequency was approximately 1.25 Hz and the filtering bandwidth was set to 0.5 Hz, which corresponds to the narrowest filtering bandwidth and which operates at the highest level of wavelet decomposition, *i.e.*, LoD = 3. At this instance, the order of the filter is maximal and the time delay corresponds to 0.39 s. In comparison, in the right inset, it can be seen that the signal frequency content is approximately 4 Hz and the filtering bandwidth corresponds to 3 Hz. In this instance, the time delay is 0.36 s, because the order of the filter is lower. In addition, it can be seen that a better attenuation is achieved, while the signal frequency is higher.

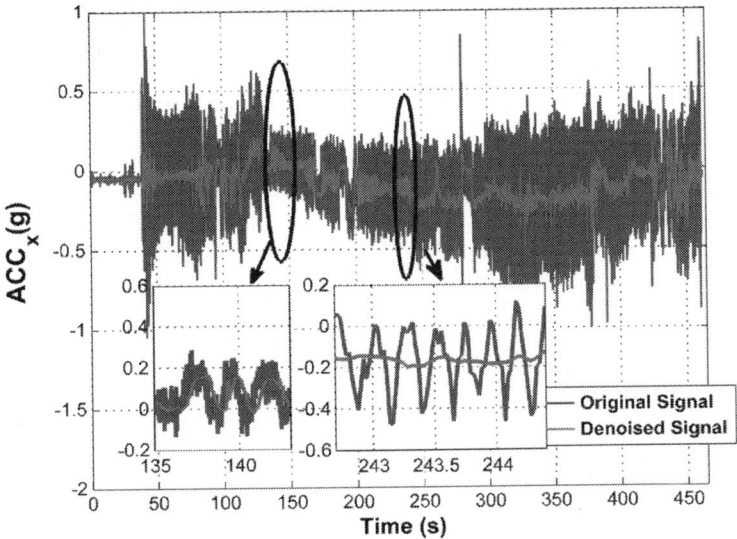

Figure 11. Variable bandwidth filtering of ACC in the x-axis.

Nevertheless, the variable filtering of ACC signals is applied only when the character of the signal content is periodic; otherwise, a constant 5.5-Hz bandwidth filter is applied the same way as used in the ARS's signals. This approach provides the same delay on both ARS and ACC data during the majority of the flight, and when low frequency vibration content occurs in ACC data, which is generally under special and rare conditions, such as the engine RPM suppression, the bandwidth of ACC data filtering is modified. This case leads to different delays for the ARS and ACC data; however, as long as they are used only for attitude compensation under steady-state conditions, these differences can be simply managed by taking different delays in data fusion into account.

4. CONCLUSIONS

This paper proposed a novel concept of filtering inertial data with an enhanced capability of providing smooth data under harsh environments, eliminating low frequency vibration influences. Cost-effective attitude and heading reference systems (AHRSs) generally fuse data from angular rate sensors (ARSs) and accelerometers (ACCs) to provide stable attitude estimation. Commonly, it is advantageous to fuse data in such a way that very low-frequency content corresponding to the steady-state flight conditions is taken from the ACC's measurements and higher-frequency content corresponding to changes of flight conditions is obtained from the ARS's measurements. Nevertheless, a problem arises when ACC readings are affected by low-frequency vibrations, and thus, the compensation ability in the attitude estimation process becomes vibration dependent. This is often the case when the correct filtering is not applied on the ACC's signals. The contribution of this paper thus lies in proposing the filtering of the ACC's with an adaptive bandwidth capability, providing the same delay in inertial data processing, when vibration frequencies are above 10 Hz. When the vibration frequency in the ACC's data is below 10 Hz, the data are filtered with a modified bandwidth to reduce the vibration effects. The modification of the filtering bandwidth relies on continuous estimation of the frequency of the strongest vibration content based on the particular bandwidth filtering applied. This filtering approach was confirmed based on simulated and real-flight data, and in all cases, the proposed approach reached better efficiency for vibration impact reduction, while preserving shorter processing delays compared with the commonly used approaches used in the commercially available AHRS systems. This paper presents the data pre-processing in terms of data filtering and not data fusion. Nevertheless, based on the effectiveness of vibration impact reduction, the proposed approach improves the ACC-based attitude compensation capability even under strong vibration, which brings a significant advantage compared with the commercially available systems.

ACKNOWLEDGMENTS

This research has been partially supported by the Technology Agency of Czech Republic (TACR) under the grant No. TA02011092 and project named "Research and development of technologies for radiolocation mapping and navigation systems", and partially by the Czech Technical University in Prague under internal grant No. SGS13/144/OHK3/2T/13.

AUTHOR CONTRIBUTIONS

The presented work contains equal contribution from the all the authors. J. Rohac defined the research theme and was responsible for the overall guidance and results analysis. M. Alam was responsible for the conceptualization of the proposed methods, filtering algorithms including the mathematical modeling.

REFERENCES

1. Gebre-Egziabher, D.; Hayward, R.C.; Powell, J.D. A low-cost GPS/inertial attitude heading reference system (AHRS) for general aviation applications. Proceedings of the IEEE 1998 Position Location and Navigation Symposium, Palm Springs, CA, USA, 20–23 April 1998.

2. Savage, P.G. Strapdown inertial navigation integration algorithm design part 1: Attitude algorithms. *J. Guid. Control Dyn.* **1998**, *21*, 19–28.

3. Savage, P.G. Strapdown inertial navigation integration algorithm design part 2: Velocity and position algorithms. *J. Guid. Control Dyn.* **1998**, *21*, 208–221.

4. Sipos, M.; Paces, P.; Reinstein, M.; Rohac, J. Flight attitude track reconstruction using two AHRS units under laboratory conditions. Proceedings of the 2009 IEEE Sensors, Christchurch, New Zealand, 25–28 October 2009.

5. Euston, M.; Coote, P.; Mahony, R.; Jonghyuk, K. A complementary filter for attitude estimation of a fixed-wing UAV. Proceedings of the International Conference on Intelligent Robots and Systems, Nice, France, 22–26 September 2008.

6. Bristeau, P.-J.; Petit, N. *Navigation System for Ground Vehicles using Temporally*; Proceedings of the American Control Conference, O'Farrell Street, San Francisco, CA, USA, 29 June–1 July 2011.

7. Reinstein, M.; Kubelka, V. Complementary filtering approach to orientation estimation using inertial sensors only. Proceedings of the IEEE International Conference on Robotics and Automation (ICRA), Prague, Czech Republic, 14–18 May 2012.

8. Farrell, J. *Aided Navigation: GPS with High Rate Sensors*; McGraw Hill Education: New York, NY, USA, 2008.

9. Simanek, J.; Reinstein, M.; Kubelka, V. Evaluation of the EKF-Based Estimation Architectures for Data Fusion in Mobile Robots. *IEEE-Asme T. Mech.* **2014**, *20*, 985–990.

10. Moorhouse, D.J.; Woodcock, R.J. *Background Information and User Guide for MIL-F-8785C, Military Specification-Flying Qualities of Piloted Airplanes*; Air Force Wright Aeronautical Labs Wright-Patterson Air Force Base: Dayton, OH, USA, 1982.

11. Pratt, R.W. *Flight Control Systems: Practical Issues in Design and Implementation*; The Institute of Electrical Engineers: Stevenage, UK, 2000; pp. 119–167.

12. Harris, F. Fixed length FIR filters with continuously variable bandwidth. Proceedings of the 1st International Conference on Wireless Communication, Vehicular Technology, Information Theory and Aerospace & Electronic Systems Technology, Aalborg, Denmark, 17–20 May 2009.

13. Harris, F.; Lowdermilk, W. Implementation Considerations and Performance Comparison of Variable Bandwidth FIR Filter and Phase Equalized IIR Filter. Proceedings of the Conference Record of the Forty-First Asilomar Conference on Signals, Systems and Computers, Pacific Grove, CA, USA, 4–7 November 2007.

14. IEEE Standard for Digitizing Waveform Recorders, 1994. pp. 1–74. Available online: http://ieeexplore.ieee.org/stamp/stamp.jsp?arnumber=469117 (accessed on 30 January 2015).

15. IEEE Standard for Terminology and Test. Methods for Analog-To-Digital Converters, 2001. pp. 1–92. Available online:http://ieeexplore.ieee.org/stamp/stamp.jsp?arnumber=929859 (accessed on 30 January 2015).

16. Kay, S.M. Detection Theory. In *Fundamentals of statistical signal processing*; Prentice Hall: Upper Saddle River, NJ, USA, 1998; Volume 2.

17. Rohac, J.; Reinstein, M.; Draxler, K. Data processing of inertial sensors in strong-vibration environment. Proceedings of the 2011 IEEE 6th International Conference on Intelligent Data Acquisition and Advanced Computing Systems (IDAACS).

18. Rohac, J.; Daado, S. Impact of Environmental Vibration on Inertial Sensor's Output. *Sens. Transducers J.* **2013**, *24*, 19–27.

19. Fugal, D.L. *Conceptual Wavelets In Digital Signal Processing*; Space & Signals Technologies LLC: Berlin Germany, 2009.

20. Rajmic, P.; Vlach, J.; Vyoral, J. Real-time wavelet transform with overlap of signal segments. Proceedings of the International Conference on Signals and Electronic Systems, Krakow, Poland, 14–17 September 2008.

21. Weeks, M. *Digital Signal Processing Using MATLAB & Wavelets*; Jones & Bartlett Learning: Sudbury, MA , USA, 2010.

22. Wojtaszczyk, P. *A Mathematical Introduction to Wavelets*; Cambridge University Press: Cambridge, UK, 1997.

Chapter 11

Noise Reduction and Gap Filling of fAPAR Time Series Using an Adapted Local Regression Filter

Álvaro Moreno [*], Francisco Javier García-Haro, Beatriz Martínez and María Amparo Gilabert

Departament de Física de la Terra i Termodinàmica, Universitat de València, Dr. Moliner, 50. 46100 Burjassot (València), Spain

ABSTRACT

Time series of remotely sensed data are an important source of information for understanding land cover dynamics. In particular, the fraction of absorbed photosynthetic active radiation (fAPAR) is a key variable in the assessment of vegetation primary production over time. However, the fAPAR series derived from polar orbit satellites are not continuous and consistent in space and time. Filtering methods are thus required to fill in gaps and produce high-quality time series. This study proposes an adapted (iteratively reweighted) local regression filter (LOESS) and performs a benchmarking intercomparison with four popular and generally applicable smoothing methods: Double Logistic (DLOG), smoothing spline (SSP), Interpolation for Data Reconstruction (IDR) and adaptive Savitzky-Golay (ASG). This paper evaluates the main advantages and drawbacks of the considered techniques. The results have shown that ASG and the adapted LOESS perform better in recovering fAPAR time series over multiple controlled noisy scenarios. Both methods can robustly reconstruct the fAPAR trajectories, reducing the noise up to 80% in the worst simulation scenario, which might be attributed to the quality control (QC) MODIS information incorporated into these filtering algorithms, their flexibility and adaptation to the upper envelope. The adapted LOESS is particularly resistant to outliers. This method clearly outperforms the other considered methods to deal with the high presence of gaps and noise in satellite data records. The low RMSE and biases obtained with the LOESS method (|rMBE| ≲ 8%; rRMSE ≲ 20%) reveals an optimal reconstruction even in most extreme situations with long seasonal gaps. An example of application of the LOESS method to fill in invalid values in real MODIS images presenting persistent cloud and snow coverage is also shown. The LOESS approach is recommended in most remote sensing applications, such as gap-filling, cloud-replacement, and observing temporal dynamics *in situ* where rapid seasonal changes are produced.

Keywords: fAPAR; noise; MODIS; time series; filtering; interpolation; LOESS

1. INTRODUCTION

The fraction of absorbed photosynthetic active radiation (fAPAR) is commonly used in ecosystem modeling because it has an important influence on exchanges of energy, water vapor and carbon dioxide between the surface of the earth and the atmosphere. The fAPAR has been also recognized as one of the fundamental essential climate variables (ECVs) by Global Terrestrial Observing System (GTOS) and Global Climate Observing System (GCOS) [1].

Space agencies and other institutional providers currently deliver various fAPAR products at different temporal and spatial resolutions over the globe taking advantage of available satellite data. A review is found in [1] and an intercomparison of products in [2,3]. However, vegetation attributes derived from polar orbit satellites, such as the EOS-MODIS, the VGT-SPOT or the Landsat-TM, often present considerable noise and poor coverage due to inefficient cloud filtering, high aerosol load, unaccounted surface directional effects, sensor problems or retrieval algorithm failures [4]. For example, MODIS fAPAR back-up retrievals have generally lower quality mostly due to residual clouds and poor atmospheric correction [5].

A well-recognized feature of fAPAR time series is a negative bias caused by snow coverage and unfavorable atmospheric conditions (cloud, ozone, dust, and other aerosols) that generally decrease near-infrared reflectance and increase reflectance in the visible, leading to spurious drops in the data [6–9]. Bidirectional Reflectance Distribution Function (BRDF) effects, mainly caused by anisotropic reflectance behavior of surfaces, may represent an additional source of uncertainty. BRDF effects can cause also positive errors in fAPAR.

Especially during winter, persistent clouds and snow coverage can originate large gaps in time series, which are not acceptable for many ecosystem or climate models [10,11]. These disturbances greatly affect the robust monitoring of land cover and terrestrial ecosystems and algorithms for time series filtering and reconstruction of fAPAR are thus required. Gap filling methods usually rely on temporal [12], spatial interpolation [13] or both [14]. Spatial interpolation has the advantage of minimizing local anomalies so that large-scale trends in regional to global vegetation phenologies can be better identified [15]. Nevertheless, spatial filters may fail when considering complex landscapes, with a high proportion of mixed pixels, where fAPAR could vary widely within a short distance [16]. This study has focused only on temporal methods.

Noise reduction and gap filling in time series is not straightforward. Filtering methods can be categorized as local or global in nature [17]. Local methods include: threshold (Interpolation for Data Reconstruction (IDR)), rank based (e.g., median, maximum or minimum filters) and linear/polynomial (e.g., moving average, adaptive Savitzky-Golay (ASG), locally weighted scatterplot smoothing (LOESS), Whittaker smoother, CACAO method, and smoothing spline (SSP) methods [18–24]). An advantage of the local methods is that they make no assumptions about the underlying nature of the processes responsible for the variability in the time series while taking advantage of the neighboring observation to generate an estimate. Nevertheless, for noisy and incomplete time series these methods are more limited and it becomes necessary to apply restrictions [25].

Global approaches consider model fits such as polynomial, Double Logistic (DLOG), asymmetric Gaussian (AG) [25–27], decomposition techniques based on Fourier analysis [28,29] or wavelet transforms [30,31]. Global methods are more suited to derive terrestrial biophysical parameters and extract seasonality information for phenological studies [25,32,33]. Nevertheless, these methods also suffer several drawbacks that limit their use. For example, the ASG filter outperformed global noise-reduction techniques when evaluated on the basis of RMSE [16]. Global methods assume an *a priori* phenological shape or magnitude that could make them not flexible enough to model the temporal response of irregular or asymmetric time series (e.g., high inter-annual variability) [19].

In the light of the abovementioned drawbacks, the present work is aimed to develop a robust methodology to provide temporally smoothed and spatially complete fAPAR time series. The study proposes an adapted LOESS to deal with the presence of noise and gaps in satellite data. The proposed LOESS method was enhanced based on an iterative reweighted algorithm to force the fAPAR curve to fit the upper envelope, thus removing contamination caused primarily by cloud and poor atmospheric correction. It is compared with four methods: DLOG, SSP, IDR and ASG, which have provided satisfying results. Hird and McDermid [34] showed that DLOG and AG techniques provide balanced performance to derive optimal times series metrics. Both methods are more flexible and effective than Fourier-based fitting method. The IDR technique compared favorably with DLOG and other smoothing techniques in [18]. The

SSP method revealed a good performance in experiments involving different data sources [35]. The ASG performed better in most situations when it was compared with other SAVER filtering approaches such as DLOG, IDR among others [36]. Atkinson *et al.* [37] showed the higher performance of the Whittaker and Fourier approaches in most cases. This study also revealed that DLOG and AG did not perform well for areas with more than one growing season per year. The DLOG and ASG techniques were successfully applied to multi-temporal NDVI data sets over Africa and northern Europe, respectively, and both are easily implemented using TIMESAT software (http://www.natgeo.lu.se/personal/Lars.Eklundh/TIMESAT/timesat.html) [25]. These methods have been used extensively on MODIS data in studies related to vegetation properties such as LAI [38] and vegetation indices [32]. The above analyses are limited and inconclusive, therefore there are still some aspects that need to be improved and completed.

This work shows the results of several experiments, which were conducted on simulated and real data sets in order to assess the performance of four methods to produce smoothed and gap-filled fAPAR data. Eight-day fAPAR time series were simulated with varying amounts of noise, and by adding gaps associated to persistent clouds and snow coverage. Typical profiles were extracted from MODIS time series that reflect the main seasonal behaviors and environmental disturbances. The applicability of the algorithm/methods on real MODIS imagery is also demonstrated.

2. METHODS AND DATA

2.1. fAPAR Estimation

The 8-day fAPAR time series have been derived from MODIS data using the algorithm proposed by Roujean and Bréon [39]. The algorithm was previously adopted in Land-SAF (Satellite Application Facility on Land Surface Analysis) to deliver operationally fAPAR from SEVIRI sensor [4] (http://landsaf.meteo.pt). Several works [40–43] have demonstrated the reliability of the proposed algorithm for retrieving daily fAPAR from medium and low-resolution satellite observations.

According to this algorithm [39], the fAPAR is obtained from the Renormalized Difference Vegetation Index (RDVI) as:

$$\text{fAPAR} = 1.81 \, \text{RDVI} - 0.21 \tag{1}$$

where

$$\text{RDVI} = (\text{NIR} - \text{RED}) / (\text{NIR} + \text{RED})^{1/2} \tag{2}$$

The reflectance in near infrared and red channels (NIR and RED, respectively) is computed in an optimal angular geometry in the solar principal plane (θ_s = 60°, θ_v = 45°, ϕ = 0°) from the BRDF parameters (k_0, k_1, k_2).

The MODIS BRDF product (MCD43A1) used to estimate fAPAR is produced every eight days by using all high quality observations acquired by both Terra and Aqua over a 16 day period [43]. MODIS data from January 2001 to December 2012 were downloaded from https://lpdaac.usgs.gov/. The product is released in a global 500 m SIN Grid. The corresponding data quality product (MC43A2) is also used in this study: layer BRDF_Albedo_Band_Quality is used to identify the best quality data for the RED and NIR bands; layer Snow_BRDF_Albedo is used to choose snow free pixels; the land/water mask comes from the layer BRDF_Albedo_Ancillary.

To make the best use of the MODIS QC information, a simplified measure of the fAPAR quality was obtained for each pixel and date. For each individual band, the QC values corresponding to the combinations of the four first bits were considered (QC = 0 for the best quality; QC = 1 for good quality; QC = 2 for low quality when more than 7 observations are used in the model inversion; QC = 3 for the worst quality when between 3 and 7 observations are considered). Quality measurements were first computed for all the six input channels (i.e., three BRDF parameter (k_0, k_1, k_2) in two MODIS bands (RED, NIR)) and then aggregated into a single quality control (QC) considering only the QC information of the bands used to compute fAPAR. Finally this ancillary information ranges from 0 (highest quality) to 6 (lowest quality).

2.2. The Proposed LOESS Method

The proposed adaptation of the LOESS method has been designed for reducing noise and suppress disturbances while maintaining the relevant data integrity. A flowchart with the adapted LOESS filtering approach is shown in Figure 1. LOESS is a locally weighted scatter plot smoother [21]. For each data value, y_i, with associate time t_i, i = 1, 2,···, N, a weighted linear least squares regression is performed to fit a d-th degree polynomial $f(t)$ = $c_0 + c_1 t + c_2 t^2 + \cdots + c_d t^d$ to all 2n + 1 points in the moving window. Each smoothed y_i^* is determined from the value of the polynomial at position t_i in a window of size 2n+1 (being n the half-width of the filtering window size).

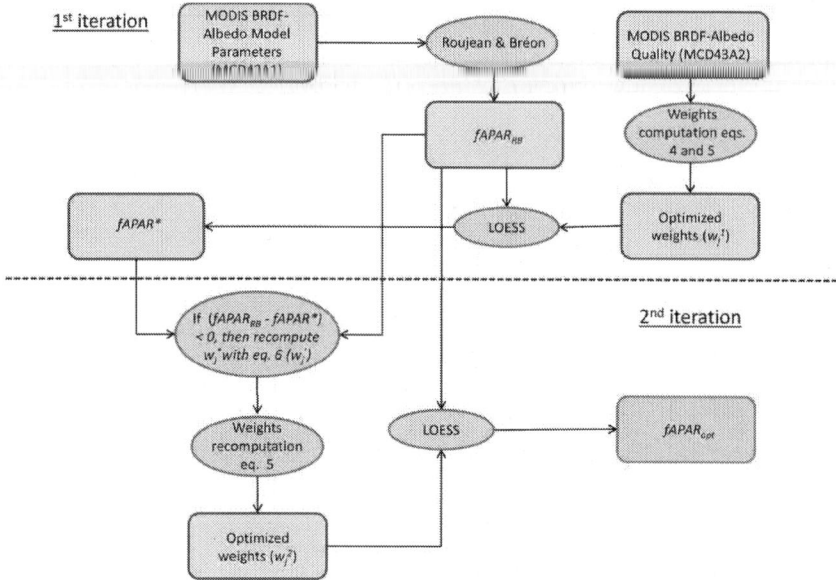

Figure 1. Flowchart of the adapted LOESS algorithm for deriving noise and gap free time series of remotely sensed fraction of absorbed photosynthetic active radiation (fAPAR).

Similarly to previous studies [25,38], we have considered the characteristics of the satellite QC information to derive an improved fAPAR filtering. Observations from optimal MODIS reflectance (QC = 0) are initially assigned with a maximum quality score (w_i = 1), invalid observations are discarded (w_i = 0), and observations with a relatively low quality are assigned with a small weight, according to the following expression:

$$w_i^* = \frac{1}{0.5QC_i + 1} \tag{3}$$

where w_i^* is the weight associated to the i-th observation and QC_i its corresponding quality flag score, which ranges from 0 to 6 (see Section 2.1).

In addition to the QC information, a robust regression weight was defined:

$$w_j = w_j^* \left(1 - \frac{|t_i - t_j|}{\Delta(t)}\right) \tag{4}$$

where t_i is the time value associated to the value to be smoothed, t_j are the neighbors defined by the moving window and $\Delta(t)$ is the half-width of filtering window. The effect of this equation is that the data point to be smoothed has the largest weight and is thus the most influential on the fit.

$$w'_j = w^*_j \left(\cfrac{1}{1 + \cfrac{|\Delta y_j|}{S\sigma}} \right) \quad \text{if } \Delta y_j < 0 \qquad (5)$$

In the above equation, Δy_j is the difference between the first fitting data (y^*_j) and the original fAPAR data (y_j); σ is the standard deviation of Δy_j and S is a free parameter (0.1 in this work) to be adjusted in order to correctly capture the upper envelope. To account for the negatively biased noise, the fitting is done in two steps.

The first step computes the weights only considering information provided by Equations (3) and (4). This allows considering QC information and increasing importance to the closest temporal steps to the smoothed point respectively. As aforementioned, the upper envelope of the fAPAR is more representative of the noise free data. To better capture it, in a second iteration of the process, the weights of each point are changed according to its residual $(\Delta y_j = y_j - y^*_j)$. Thus, points of the original fAPAR data lower that the smoothed ones are assumed to be far from the upper envelop and assigned with lower weight according to Equation (5).

2.3. Methods under Comparison

In the DLOG method [25,27], a double logistic (asymmetric) function is fitted to data in intervals around maxima and minima in the time-series. The local model function has two linear parameters determining the base level and the amplitude and four non-linear parameters, which determine the shape of the basis function. DLOG implementation in TIMESAT uses a Levenberg-Marquardt algorithm [44] to solve non-linear least squares minimization problem.

The SSP method does a piecewise smooth (cubic polynomial) curve fitting for a time series in such a way that the first and second derivatives of the resulting curve are continuous. This method is adjustable through a single smoothing parameter λ, which controls the smoothness of the cubic spline. In the limit $\lambda = 0$, no departures of the data points from the fitted curve are found, whereas for high values of λ, it controls the trade-off between fidelity to the data and roughness of the function estimate. SSP is similar to the widely used Whittaker smoother [22] but without any adaptation to deal with the negatively biased noise in the time series.

The IDR filter [18] creates an alternative time series by computing the mean of the immediate preceding and following observations. The filtered value replaces the original value when the difference is above a certain threshold.

The ASG [19] filter is a generalized moving average, which preserves better the features of the underlying data value and its derivatives. The method is fast since coefficients are computed in advance [45] using an unweighted linear polynomial of a given degree, d. A quadratic polynomial ($d = 2$) is usually assumed (e.g., TIMESAT [25]). The main disadvantage of the method is that it requires a series of equally spaced data points.

2.4. Experiment Settings

In this work, the gaps in the time series with DLOG, SSP, IDR, and ASG were filled with a linear interpolation method. In case of the LOESS method, an adaption of the weights assignment to deal with missing data has been also considered. Thus, when no data was available a zero weight is assigned to that point and it is automatically interpolated.

To obtain the optimum parameters for each of the other four approaches, different fitting criteria were set to reconstruct the fAPAR time series over the nine reference sites in the study area. The selected methods were applied as close as possible to their standard implementation including the original parameterization as proposed by their authors. The SSP parameter was evaluated using the IMSL (International Mathematics and Statistics Library) routine CSSMOOTH, as implemented in the IDL (Interactive Data Language) environment using a Cross Validation procedure [46]. A similar implementation was evaluated in [35]. Although the optimum smoothing parameter λ for the spline curve was derived automatically, the goodness of the fit was confirmed by visual inspection. IDR was threshold to 0.02, as proposed by the authors [18] and used in recent studies [36]. Only a marginal improvement was found by a fine tuning of this parameter.

The ASG and DLOG approaches were adjusted interactively for each site in TIMESAT software to arrive at close fitting results. This manual adjustment has been recommended by the TIMESAT authors [47] and adopted in previous works [16,19].The six parameters of the DLOG model were obtained automatically by TIMESAT using the Levenberg-Marquardt method [44]. Initial values of the non-linear parameters are obtained by looping through a number of pre-defined model functions in a highly efficient search routine.

Both LOESS and ASG require specifying two parameters, the half-width of the filtering window size, n, and the degree of polynomial, d. The value of d determines the degree of filtering, but it also affects the ability to follow rapid changes [48]. In general, high d values may overfit to the data and give less smooth results [19]. For LOESS, if there are enough high quality data in the temporal window, a five-order polynomial ($d = 5$) was chosen for the weighted fit. Under extreme conditions, high percentage of occurrence of missing data (bad conditioned problem), LOESS adjusts a simpler curve (linear) to the sparsely available high-quality observations of the gap. Furthermore, to avoid to deal with such extreme conditions, we adopted a wide value of the half-width temporal window, *i.e.*, $n = 8$. For ASG, the values $d = 2$ and $n = 5$. Chen *et al.* [17] recommended as appropriate parameters n in the range 4–7 and d in the range 2–4 ($d = 2$ is fixed by TIMESAT).

Using these optimal parameters, the five approaches were applied on different simulated and actual fAPAR time series. To assess the influence of the analyst subjectivity in the "tuning" step of each method, the same adjustment was achieved by three different analysts. Small differences were obtained for the considered statistical indicators.

2.5. Simulation of fAPAR Time Series

Typical fAPAR profiles were obtained based on 12 years (2001–2012) of actual fAPAR time series for a selection of representative sites [23,34]. Many different test scenarios were analyzed considering different levels of noise and missing data. The main steps of this procedure are:

(1) *Selection of representative sites:* Nine different homogeneous sites have been selected covering very different vegetation types in the study area (see Table 1). They were identified from a hybrid land-cover map by the synergistic combination of four classifications: CORINE, GLC2000, MODIS and GlobCover [49]. To ensure a good homogeneity and reduce the impact of misregistration errors, we imposed the agreement between the pixel label and modal land-cover label within a 3 × 3 pixel window (with a frequency of modal value higher than 6). We additionally imposed neighborhood pixel homogeneity of fAPAR time profiles (*i.e.*, selecting maximum allowable variance values). For each site, 12-year periodic time profiles were obtained in a two-step process. First, a typical yearly noise free fAPAR profile was calculated for each site by averaging the pixel fAPAR values with the highest quality (QC = 0). Second, these annual profiles were replicated for 12 times in order to obtain statistically significant results.

Table 1. List of the nine sites selected for the study.

Site	Latitude	Longitude	Land Cover Type
1	37°47'41"N	1°32'9"W	Irrigated crops
2	40°40'11"N	0°24'7"E	Non irrigated crops
3	38°51'58"N	2°54'39"W	Mosaic of croplands
4	37°6'26"N	6°40'11"W	Broadleaved forest
5	41°17'41"N	0°42'20"E	Needleleaved forest
6	41°14'28"N	4°29'28"W	Mixed forest
7	39°10'11"N	0°46'37"W	Shrublands
8	41°23'34"N	0°47'41"E	Grasslands
9	41°4'49"N	0°48'45"W	Sparse vegetation

(2) *Addition of noise component and missing/invalid data.* The underlying time series was subsequently perturbed by adding noise with varying levels. Noise components were generated using a random normal distribution $N(\mu,\sigma)$. The values for μ and σ were estimated using the range of noise found in the study area corresponding to different levels of quality in the data (high, medium, poor). In all cases, negative μ values were used, reflecting the perturbation in fAPAR time series due to cloud and poor atmospheric conditions. The noise was added to a uniformly random sample of points, whereas the rest of data was noise-free. To evaluate the impact of invalid values (gaps) in the reconstruction, two different types of gaps distributions were evaluated: (i) a random uniform distribution and (ii) a distribution of missing data extracted from real QC information of MODIS BRDF time series for selected sites. The simulation of typically adverse environmental conditions (e.g., persistent snow/cloud during winter time), which may lead to 2–4 months with consecutive missing data, is very interesting for our purposes.

2.6. Reconstruction Accuracy

The performance of the five filtering methods was quantified in terms of the mean absolute error (MAE) and root mean-squared-error (RMSE) as a measure of the accuracy, and the mean error (MBE) as a measure of the bias:

$$\text{MAE} = \frac{1}{N} \sum_{i=1}^{N} |y_i^* - y_i| \tag{6}$$

$$\text{RMSE} = \frac{1}{N} \sqrt{\sum_{i=1}^{N} (y_i^* - y_i)^2} \tag{7}$$

$$\text{MBE} = \frac{1}{N} \sum_{i=1}^{N} (y_i^* - y_i) \tag{8}$$

where y^* is the filtered fAPAR, y is its true value and N is the number of points in the time series. These statistics were evaluated in noise-controlled scenarios over typical profiles of fAPAR created with real satellite data (MODIS). In addition, they were also computed for the simulated (unfiltered) time series. This allowed us to determine a percentage of filtering improvement according to the following expressions:

$$rMBE(\%) = 100\% \frac{MBE_{filtered}}{MBE_{no\ filtered}} \qquad (9)$$

$$rMAE(\%) = 100\% \frac{MAE_{filtered}}{MAE_{no\ filtered}} \qquad (10)$$

$$rRMSE(\%) = 100\% \frac{RMSE_{filtered}}{RMSE_{no\ filtered}} \qquad (11)$$

3. RESULTS

3.1. Scenario 1: Different Levels of Noise in 25% of the Data

The first simulation scenario consisted of adding a noise distribution to 25% (random uniform) of the data points. 25 artificial datasets were thus generated with noise mean μ ranging from −0.01 to −0.05 (step = −0.01) and noise standard deviation σ ranging from 0.01 to 0.05 (step = 0.01). Figure 2 shows an example of fAPAR reconstruction under very noisy conditions corresponding to mosaic croplands, composed of small patches with different crop types. Four of the methods (except for the SSP) are designed for adaptation to the upper envelope. At giving priority to high values, this might cause biases toward relatively high values and overlook some useful information contained in low values. In particular, the high frequency of positive peaks remains in the case of the IDR method, at preserving the maximum value within its short temporal window. This method, although partially filters out the introduced noise, still produces shaky profiles with misleading peaks. The SSP, LOESS and ASG methods show to be effective in removing the high frequency noise while reproducing accurately the expected seasonal cycle. The cubic splines produce a remarkable smooth curve although its main drawbacks is that it does not preserve well some temporal metrics such as the amplitude of the annual fAPAR cycle and the rate of increase/decrease during the development and senescence periods.

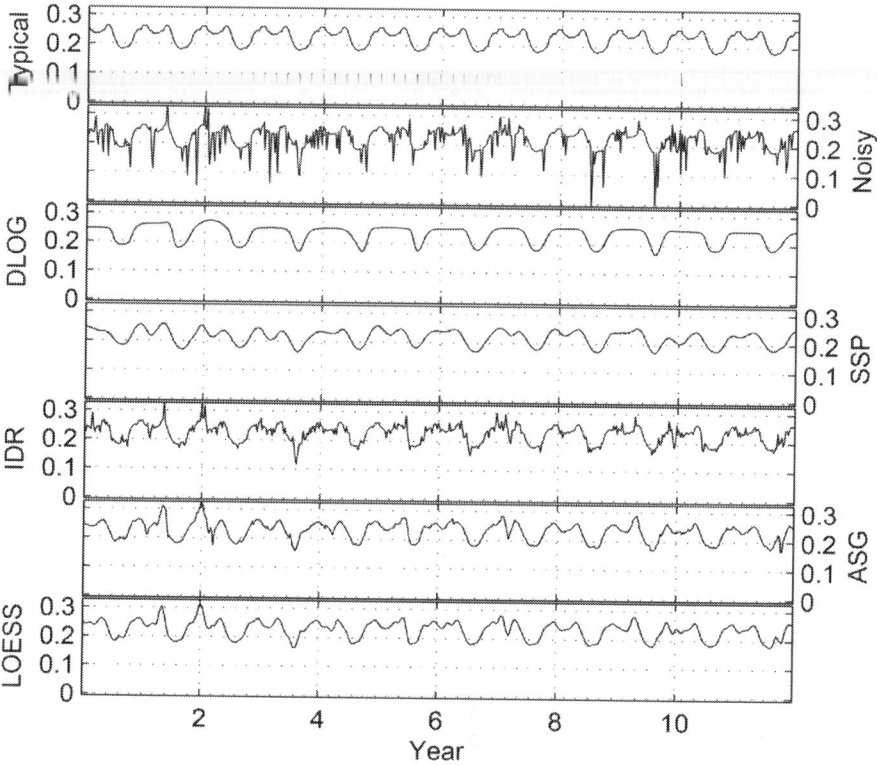

Figure 2. Reconstructed fAPAR series corresponding to a mosaic of croplands area. The original fAPAR (**top**) is followed by the noisiest scenario with $N(\mu = -0.05, \sigma = 0.05)$ in 25% of the points and the four filtered signatures.

Tables 2 and 3 present the average values of the performance measures over all the selected sites and methods. The LOESS and ASG are clearly more accurate (MAE, RMSE) and unbiased (MBE) than the rest of methods. In relative terms, both methods are effective in reducing both bias (|rMBE| << 100) and inaccuracy (rRMSE<<100).Where no distinct negative bias is present ASG filter performs better than others independently of the noise variance, while LOESS would be more properly applied *in situ* of negatively biased noise (a more realistic case). Conversely, the DLOG rather than enhancing the quality of the time series increases both types of errors when lowest bias and error magnitude are considered. The SSP method preserves the original bias (*i.e.*, |rMBE| = 100) in all cases as expected since it is not adapted to correct negative biases due to clouds and poor atmospheric correction. Finally, the effectiveness of IDR to reconstruct the original timeseries is small, e.g., a marginal improvement in accuracy is found for simulations with $\mu = -0.01$, $\sigma = 0.01$. It is important to note that the IDR shows an improved performance for the worst noise scenario considered in this study ($\mu = -0.05$, $\sigma = 0.05$).

Table 2. Performance measures averaged over the nine sites, for simulations with $\mu = -0.01$, $\sigma = 0.01$.

	MBE	MAE	RMSE	rMBE(%)	rMAE(%)	rRMSE(%)
DLOG	0.005	0.009	0.012	-300	300	180
Spline	-0.002	0.003	0.004	100	120	60
IDR	-0.0018	0.003	0.006	90	90	90
ASG	0.0018	0.003	0.004	-80	90	50
LOESS	0.0008	0.0017	0.003	-30	60	40

Table 3. Performance measures averaged over the nine sites, for simulations with $\mu = -0.05$, $\sigma = 0.05$.

	MBE	MAE	RMSE	rMBE(%)	rMAE(%)	rRMSE(%)
DLOG	0.006	0.009	0.012	-60	60	37
SSP	-0.009	0.011	0.014	100	80	40
IDR	-0.0014	0.006	0.014	13	40	40
ASG	0.0016	0.003	0.004	-16	17	11
LOESS	0.0010	0.003	0.004	-10	18	10

The influence of the noise parameters reveals that ASG and LOESS clearly outperform the other methods (see example in Figure 3). ASG performs slightly better than LOESS although LOESS exhibits also very high accuracies, with RMSE typically lower than 0.005. Overall, the reconstruction based on the DLOG algorithm generates the highest RMSE values because the method fails to capture components presenting multiple vegetation cycles per year (e.g., presence of a double phenology) in the original signals. The accuracy of this method becomes rather insensitive to the noise parameters. The performance of IDR and SSP approaches is also poor. The bias is the factor that most influences the accuracy of SSP, whereas the IDR is mainly influenced by the standard deviation σ. In general, the accuracy of IDR, ASG and LOESS increases considerably when noise mean values are far from zero since these methods rely on the assumption that the atmospheric contamination tends to drop fAPAR values, and therefore, higher fAPAR values are more acceptable than lower ones.

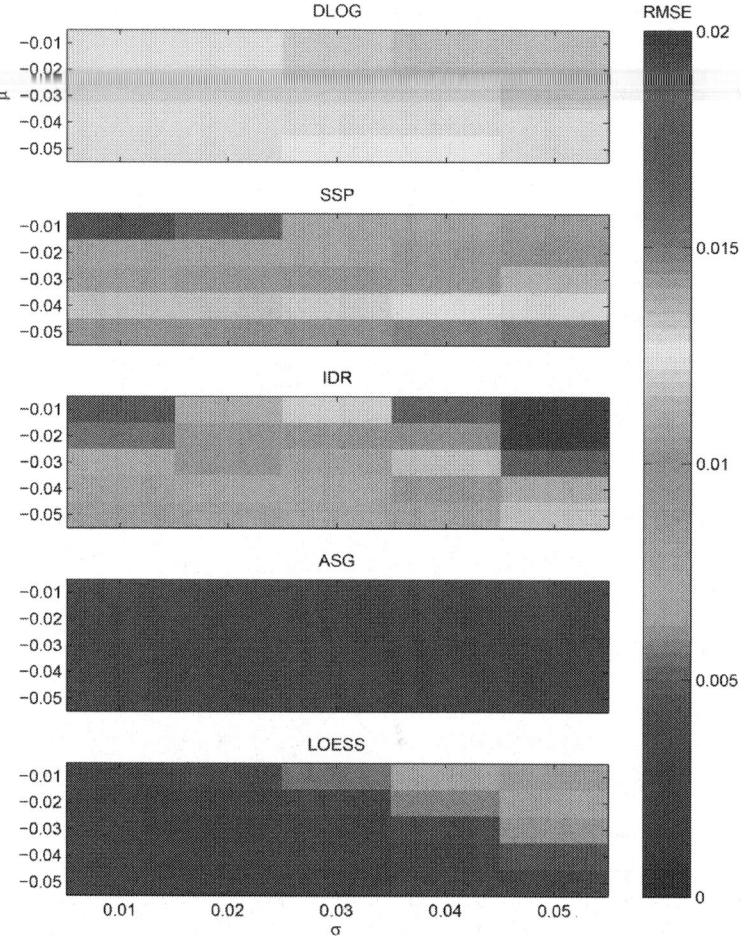

Figure 3. Comparison of the RMSE for the 25 $N(\mu, \sigma)$ simulations of the five filtering methods in a mixed forest area.

3.2. Scenario 2: Variation in the Proportion of Noisy Data

The second simulation scenario consisted of fixing the parameters describing the noise distribution ($\mu = -0.025$, $\sigma = 0.025$) and varying the percentage of randomly distributed samples contaminated by noise. Figure 4 shows that the DLOG method produces fAPAR with a significant positive bias (MBE ~ 0.004). Conversely, SSP and to a lesser extent IDR produce a high negative bias. This underestimation of fAPAR increases to inacceptable levels (MBE < -0.010) in proportion to the amount of data affected by noise and is related to the inability of these methods to capture the fAPAR envelope.

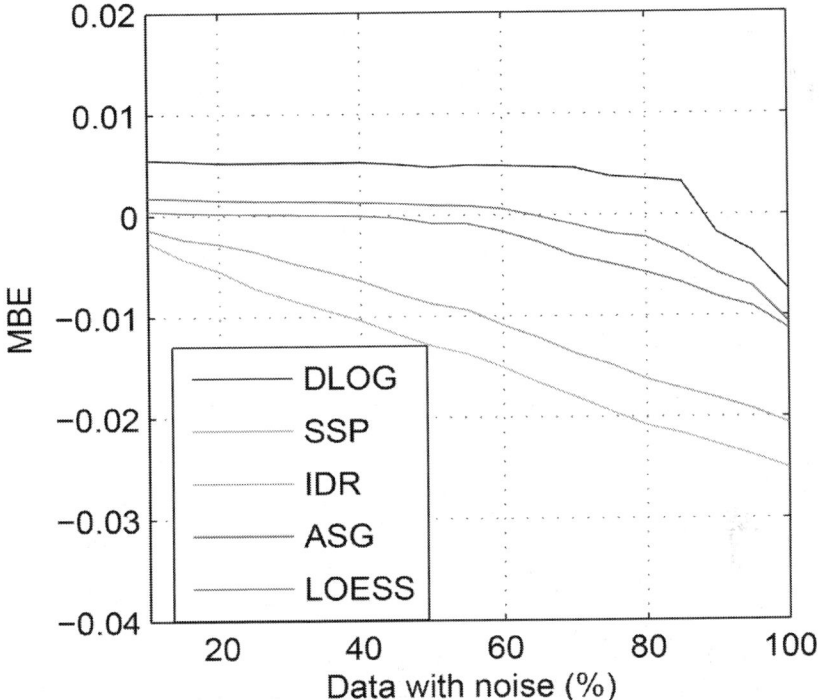

Figure 4. Performance of the applied filtering methods in terms of bias (MBE) averaged over the nine sites.

The LOESS presents generally a negligible bias (|MBE| < 0.002), but it increases leading to significant negative bias when the percentage of noisy data exceeds the 70%. When this scenario occurs, there is not an optimal method to reconstruct the upper envelope. This is partly due to the fact that the highest values are also generally affected by noise and therefore they have lower weights because of their associated QC values. The ASG method performs almost equally well and is even more robust when the time series is fully contaminated with noise especially when low biased noise is considered.

Figure 5 shows that the SSP and IDR clearly manifest a negative trend in their accuracy as the data affected by noise increases. The performance of the DLOG becomes acceptable when the percentage of contaminated data is high (>40%). However, as this situation is uncommon in the study area it difficult the practical applicability of the DLOG method.

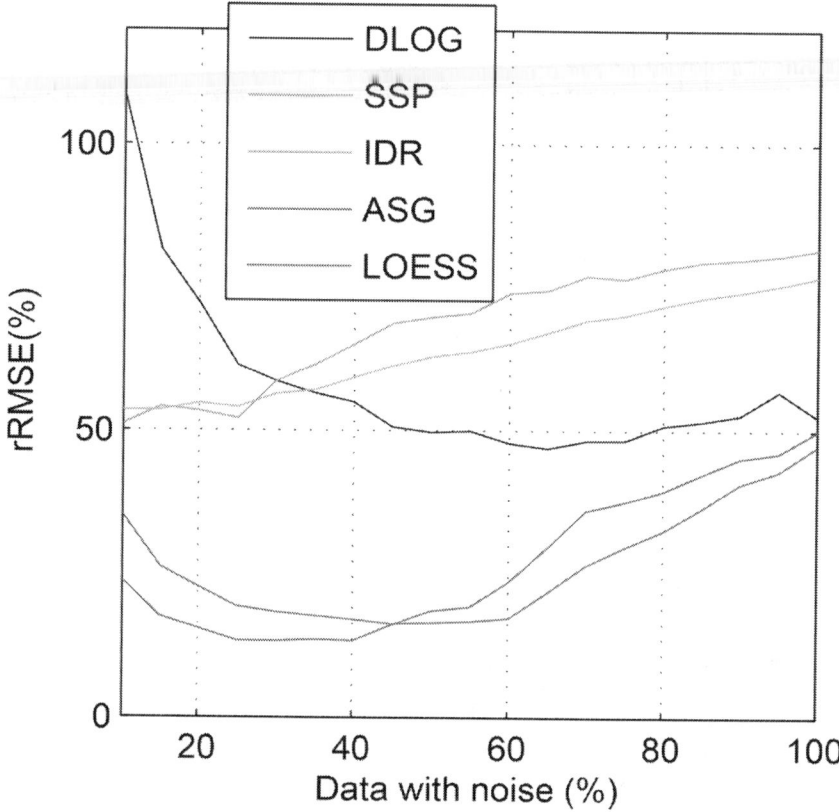

Figure 5. Performance of the applied filtering methods in terms of rRMSE averaged over the nine sites.

The performance of the other methods, ASG and LOESS, is considerably better under situations of 20%–40% of contaminated data. Notably, the LOESS method reduces the noise disturbance to less than 1/4 of the original value. Again an increase of the LOESS and ASG error is observed for high percentage of noisy data (>50%). As aforementioned, this is related with the confounding factor of having highly positive noisy values. This scenario is unrealistic (noise usually cause rapid drops in fAPAR time series) and so it has not been considered in the adaptation of the proposed LOESS approach and in ASG.

3.3. Scenario 3: Randomly Distributed Missing Data

To assess the effectiveness of the four methods in presence of invalid values (gaps), we simulated noisy data with random uniform gaps in a two-step process. It is important to note that while LOESS deals with missing data in this parameterization, the other methods require gaps to be filled before its use.

Following previous studies [16,19], a linear interpolator was used in these cases, independently of the size of the gaps. First, a set of noisy data was obtained by adding Gaussian noise ($\mu = -0.025$ and $\sigma = 0.025$) randomly to 25% of fAPAR values. Second, different data sets were generated by assuming that a subset of the data presented invalid values. The proportion of gaps ranged from 5% to 30%.

Figure 6 shows the reconstructed fAPAR series for a mixed forest from a highly noisy time series. Compared with scenarios 1 and 2 (see Figure 2) all five methods produce smoother fAPAR profiles in presence of missing data. This is partly due to the use a linear interpolator to replace missing data, thereby limiting the temporal variance and performing an additional filtering. This may destroy relevant information (especially during long periods with missing data), as it is particularly evident in the case of the DLOG method that removes the peak of the second season (spring).

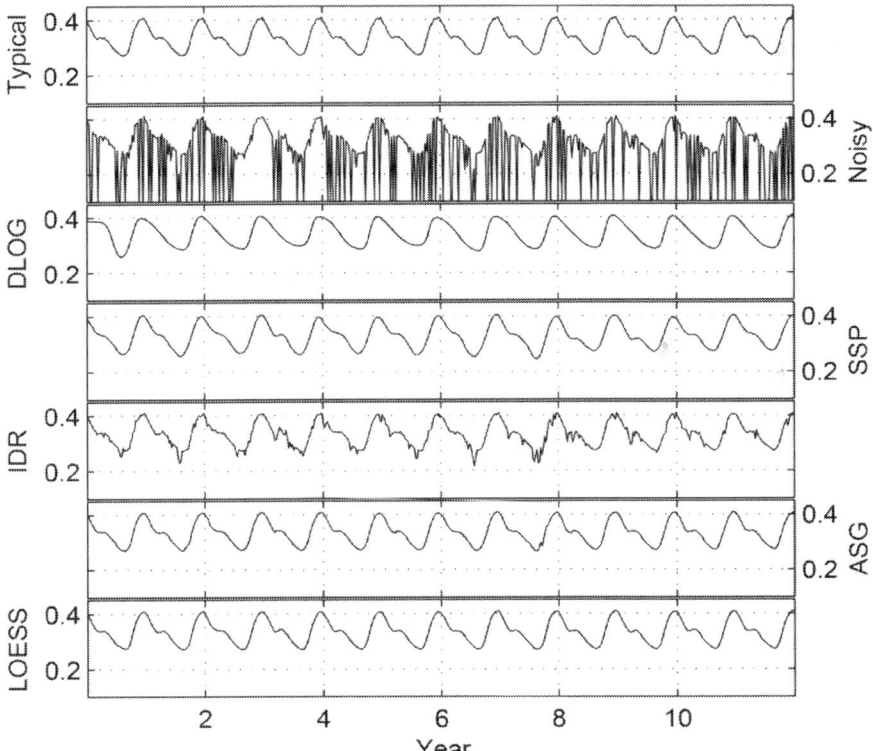

Figure 6. Reconstructed fAPAR series corresponding to a mixed forest area. The original fAPAR (top) is followed by the noisiest scenario with N ($\mu = -0.25$, $\sigma = 0.25$) in 25% of the points and a 15% of missing data for the four filtered signatures.

The reconstruction accuracy, in terms of MAE, is shown in Table 4. It is important to note that the methods are sorted according to their performance. LOESS provides the more accurate reconstruction (MAE in the range 0.0016–0.0025), followed by ASG (0.0025–0.0035), IDR (0.0044–0.0063), SSP (0.0072–0.0076) and DLOG, which provides the poorest results (0.0086–0.0088). The influence of the amount of gaps allows differentiating two different behaviors: (1) The reconstruction error of IDR, ASG and LOESS methods increases with the proportion of gaps in the series. (2) The error of DLOG and SSP seems to remain almost constant, partly because the increase in the number of gaps produces also a diminution in the number of noisy observations. In fact, both random temporal distributions are different and can be coincident, replacing noisy data with gaps.

Table 4. Average MAE (nine selected sites) with noise parameters of N (μ = −0.025, σ = 0.025) and considering varying gap fractions.

	Gap Fraction (%)					
	5	10	15	20	25	30
DLOG	0.0088	0.0088	0.0089	0.0087	0.0086	0.0088
SSP	0.0077	0.0075	0.0075	0.0072	0.0073	0.0076
IDR	0.0044	0.0046	0.0051	0.0051	0.0056	0.0063
ASG	0.0025	0.0026	0.0029	0.0030	0.0032	0.0035
LOESS	0.0016	0.0017	0.0019	0.0018	0.0020	0.0025

3.4. Scenario 4: Influence of Missing Data Extracted from a Real Distribution

Data were simulated following the same procedure as in 3.3 but using a more realistic distribution non stationary (seasonal) of the missing data, since they were extracted from QC of MODIS BRDF time series. Four distributions of gaps were selected from four different sites, as shown in Figure 7. They characterize typical situations with missing values during long, continuous periods. In particular, the winter period (December–March) shows the highest frequency of invalid observations, which can be attributed to the periodical snow cover and high cloudiness in these regions. These simulations provide a more challenging scenario for testing the reconstruction ability of the five methods.

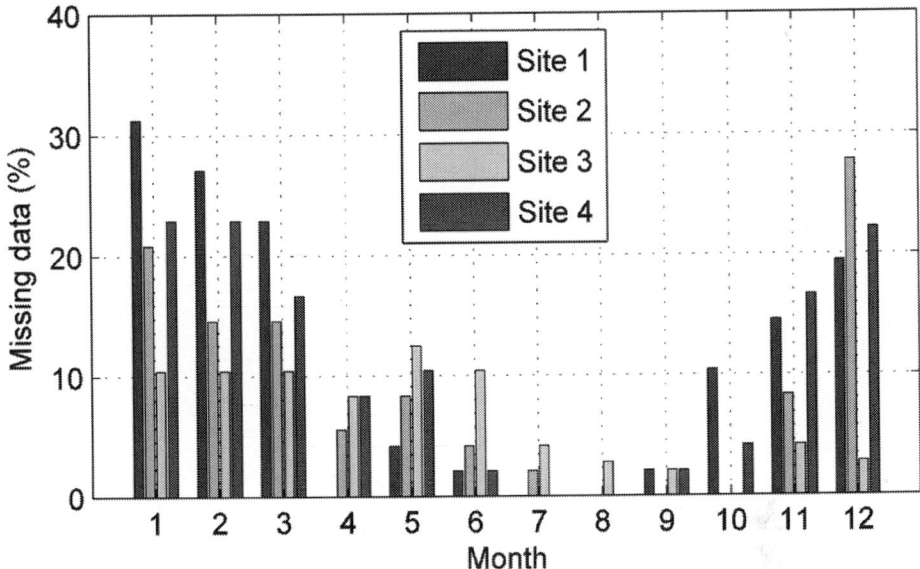

Figure 7. Monthly distribution of the gaps for the four sites considered.

From Table 5, one can observe that DLOG and SSP are the most inaccurate methods. This is particularly exacerbating for the case of the DLOG, which provides relative values of improvement exceeding 100 (*i.e.*, filtered fAPAR values depart from the "true" time series more than unfiltered ones). The worst results (highest MBE, MAE and RMSE) are found for DLOG method, although the performance of this method does not deteriorate at increasing the percentage of gaps (constant values of MAE). The SSP provides also biased fAPAR estimates since it was not designed to capture the upper envelope. The ASG turns out to be more accurate than the IDR method, providing a good fit to the original data (e.g., it reduces RMSE to less than 40%). As in previous results, the LOESS method provides the lowest biases and interpolates more precisely than the other considered methods, which can be partly due to the use of an optimal (non-linear) local interpolator. For example, the LOESS reduces more than 92% of the introduced bias (*i.e.*, |rMBE| < 8%) and moreover it produces the higher accuracy (*i.e.*, rRMSE ≈ 20%).

3.5. Application to Real Imagery

In this section, an application of the optimized LOESS algorithm in actual MCD43A1 (BRDF MODIS parameters) and MCD43A2 (BRDF QC) data has been considered to perform a qualitative analysis of the proposed methodology. Figure 8shows the reconstructed and original fAPAR profiles corresponding to four sites with different vegetation types (needleleaf and broadleaf forests, shrublands and non-irrigated crops). LOESS performs as well as expected. That is, it is able to correctly capture the upper envelope, removing the unrealistic high frequency noise and reconstructing the present gaps in the time series for all the test sites considered. In addition, no significant differences are noticeable in comparison with Figure s 2 and 6 when the synthetic noise was added to time series.

Table 5. Statistical results of the different filtering methods considered using extracted MODIS real gap profiles.

		MBE	MAE	RMSE	rMBE (%)	rMAE (%)	rRMSE (%)
Profile 1	DLOG	0.005	0.009	0.012	−60	120	70
	SSP	−0.006	0.007	0.010	100	100	50
	IDR	−0.004	0.005	0.012	60	70	70
	ASG	0.0008	0.003	0.006	−10	40	40
	LOESS	−0.0002	0.002	0.005	−6	30	19
Profile 2	DLOG	0.006	0.009	0.014	−100	130	80
	SSP	−0.006	0.007	0.010	100	100	50
	IDR	0.0003	0.005	0.012	50	70	70
	ASG	0.019	0.003	0.008	−30	50	40
	LOESS	0.0004	0.0019	0.0034	−2	30	19
Profile 3	DLOG	0.005	0.009	0.012	−80	120	70
	SSP	−0.006	0.007	0.009	100	100	50
	IDR	−0.004	0.005	0.011	60	60	60
	ASG	0.0016	0.003	0.004	−30	30	20
	LOESS	0.0004	0.0019	0.0032	−8	30	20

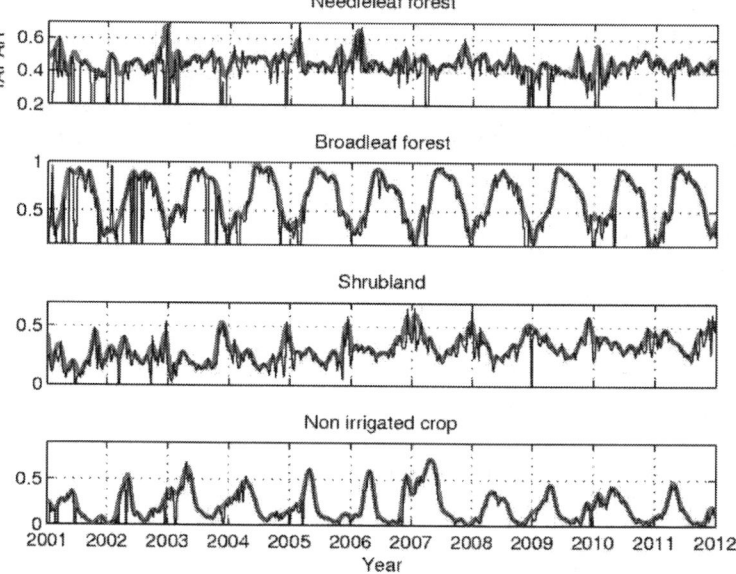

Figure 8. Application of the optimized LOESS method on real MODIS data. Black line corresponds with unfiltered fAPAR data and red line with the filtered one.

Finally, an example of the LOESS application to MODIS scale is shown in Figure 9 for the 1 January 2011 over the Iberian Peninsula. The visual comparison of the images indicates that the LOESS approach allows preserving almost identical the present spatial patterns after its application. In relation to the missing data, it is able to recover the information from neighboring dates without any visual noticeable spatial artifacts.

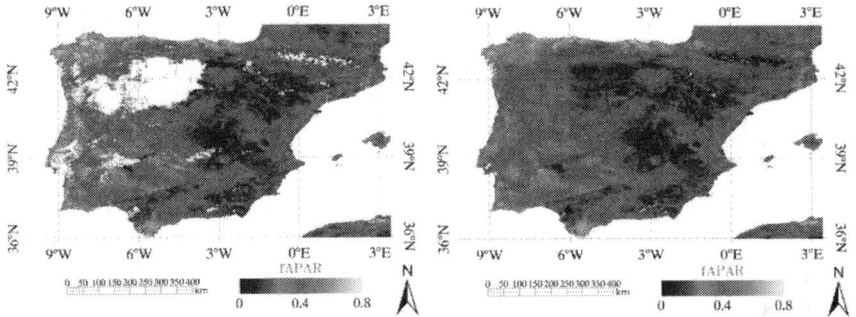

Figure 9. fAPAR values for the Iberian Peninsula before (**left**) and after (**right**) the application of the adapted LOESS method for filtering and interpolation using real MODIS data (products MCD43A1 and MCD43A2).

4. DISCUSSION

In time series analysis of fAPAR products, inaccuracies in modeling the BRDF effects do not necessarily introduce a negative bias, but they can cause also positive errors in fAPAR. In this paper, the fAPAR was estimated using an optimal angular geometry. The use of a BRDF MODIS product (MCD43A1) may lower these influences, which could be more significant when using daily MODIS reflectance products due to the different image acquisition angles. For this reason, the common assumption that nearly always disturbances in these time series are caused by cloud cover and atmospheric contamination is accepted.

One typical problem in estimating the quality of the temporal reconstruction techniques is the unavailability of an objective reference data used to validate the results. In order to solve this problem, different approximates need to be made in practice. For example, [36] evaluate the performance of each filtering technique using as a reference the mean time series (assumed to be accurate) obtained from eight noise-reduction techniques, which is a simplification. Atzberger and Eilers [22] propose a range of approaches to test the plausibility of filtering results including class discrimination, pseudo-invariant targets, geostatistical noise estimation, *etc.* We have considered as the reference synthetic data representing multiple controlled scenarios. They include Gaussian noise distributions with varying amplitude and negative bias as well as occurrence of missing data extracted from real MODIS time series. Although, this Gaussian noise is commonly used in the literature to generate artificial time series [35,37,50], a red noise was further applied providing similar results. Future work is required to implement more realistic noise patterns since cloudy periods are usually concentrated in known period of the year.

The five considered approaches are computationally inexpensive and easy to implement, which make them applicable to generate fAPAR time series for regional and global studies. The methods are also applicable to NDVI and other remote sensing biophysical products such as LAI. Although finding one single optimal value for the two LOESS parameters that can be used in any circumstance is not feasible, we recommend a value around $n = 8$ for the half-width of the filtering window size and $d = 5$ for the degree of polynomial. Low differences were observed in the filtered curves under small perturbations of these parameters. The settings of the other four algorithms followed the original parameterization, showing to be relative robust against the analyst subjectivity.

The LOESS and ASG are clearly more accurate and unbiased than the rest of methods, since they are more flexible and best suited to capture the rapid seasonal changes in vegetation. An important cause for the moderate to poor performance of the SSP is that it assigns weights that are symmetric for points above and below the curve. In contrast, the other four approaches assume that clouds and poor atmospheric conditions usually depress fAPAR values. Hence they use some iterative tuning to remove these sudden events at punishing negative deviations in the weighted fitting, while preserving those compatible with the gradual process of vegetation change. One possible limitation is that this might cause biases toward relatively high values and overlook some useful information contained in low values.

Three of the approaches under comparison (LOESS, ASG and DLOG) incorporate ancillary data in the form of prior knowledge about the quality of the inputs (i.e., MODIS QC), which is routinely provided by most remote sensing products nowadays. In general, by maximizing the use of high-quality data to replace missing or poor-quality observations in the iterative process may enhance the quality of the filtering process. Further improvement could be obtained by modifying the original IDR and SSP algorithms in order to incorporate this MODIS QC.

The capacity of the methods resilient to periods of missing data to provide realistic interpolation is challenged in cases corresponding to medium to high fraction of missing data [16]. Techniques based on the processing of the time series as a whole (DLOG) were not demonstrated to perform systematically better than techniques based on a limited temporal window (LOESS, ASG, IDR) although they were expected to fill long gaps with the data available for the several years available in the time series. This is consistent with the finding that MAE values remain almost constant for DLOG in Tables 4 and 5, showing no influence of the amount of gaps. IDR was generally more faithful but showed poor capacity to fill long gaps, e.g., to reconstruct data contaminated by clouds during more than 4–5 consecutive observations. The ASG clearly outperforms the DLOG, the IDR and the SSP methods. These findings are consistent with other literature [36,50], showing better reconstruction regarding to other common techniques. Although the ASG is effective to denoise regularly sampled observations, its effectiveness diminishes in presence of missing records, particularly for long seasonal gaps (winter snow coverage and persistent cloud occurrence). Finally, the low RMSE and biases obtained with the LOESS method (rMBE < 8%; rRMSE < 20%) reveals an optimal reconstruction even in most extreme situations with long seasonal gaps. The linear approach could be excessive under prolonged periods of missing data. The LOESS method, which relies in a more complex fitting function (based on a wider temporal window with a higher degree polynomial to address rapid changes and an optimal weighting of observations), has allowed preserving and recovering the relevant information of the fAPAR.

5. CONCLUSIONS

Satellite land products generally need to be processed to remove data gaps and low quality data caused by cloud/snow contamination or algorithm limitations before they can be used in models directly. This work has developed an iteratively reweighted LOESS approach for noise reduction and gap filling of fAPAR time series. It was enhanced based on an iterative reweighted algorithm to force the fAPAR curve to fit the upper envelope, thus removing contamination caused primarily by cloud contamination and poor atmospheric conditions. The method provides a simple parameterization that converges to the upper envelope using only two iterations.

This approach has been compared quantitatively with four approaches, the DLOG, SSP, IDR and ASG methods, widely applied to reconstruct time series of remote sensing data. Results evidence the outperformance of the LOESS and ASG methods when compared with DLOG, SSP, and IDR over multiple controlled noisy scenarios. The DLOG, IDR and SSP methods introduce significant biases. The SSP, LOESS and ASG methods show to be effective in removing the high frequency noise while reproducing accurately the expected seasonal cycle, although the SSP is not robust against cloud contamination effects and hence it is unable to recover the fAPAR trajectories with different biases of noise. The LOESS and ASG are clearly more accurate and unbiased than the rest of methods, since they are more flexible and best suited to capture the rapid seasonal changes in vegetation. The other considered methods evidence also important drawbacks. For example, the DLOG has obtained the worst performance especially when the temporal evolution of the fAPAR presents more complex patterns.

Furthermore, LOESS is clearly the most accurate gap filler (MAE < 0.0022), even *in situ* with a large fraction of non-stationary (seasonal) gaps. The LOESS method does a credible job of filling long seasonal gaps due to its nonlinear nature and the use of wider moving window with a weighting scheme. This is partly due to its unique parameterization based on a nonlinear local interpolation in conjunction with a weighting scheme that optimally takes into account the ageing of the observations.

Although the selection of the optimal smoothing method depends greatly on the intended application, LOESS is expected to have the best performance in most applications, such as gap-filling, cloud-replacement, and observing temporal dynamics*in situ* where rapid seasonal changes are produced.

ACKNOWLEDGMENTS

This research was supported by the projects: RESET CLIMATE (Spanish Ministry of Economy and Competitiveness, CGL2012-35831), ERMES (EU FP7-Space-2013, Contract 606983) and LSA SAF (EUMETSAT). Special thanks are due to the anonymous referees for their suggestions, which greatly improved the quality of the paper.

AUTHOR CONTRIBUTIONS

All authors made significant contributions to the results, interpretation and presentation of this Article. All authors read and commented on the manuscript.

REFERENCES AND NOTES

1. Gobron, N.; Verstraete, M. ECV T10: Fraction of Absorbed Photosynthetically Active Radiation (fAPAR) Essential Climate Variables. Available online: http://www.fao.org/gtos/doc/ECVs/T10/T10.pdf (accessed on 26 August 2014).

2. Meroni, M.; Atzberger, C.; Vancutsem, C.; Gobron, N.; Baret, F.; Lacaze, R.; Eerens, H.; Leo, O. Evaluation of agreement between space remote sensing SPOT-VEGETATION fAPAR time series. *IEEE Trans. Geosci. Remote Sens* **2013**, *51*, 1951–1962.

3. Martínez, B.; Camacho, F.; Verger, A.; García-Haro, F.J.; Gilabert, M.A. Inter-comparison and quality assessment of MERIS, MODIS and SEVIRI fAPAR products over the Iberian Peninsula. *Int. J. Appl. Earth Obs. Geoinf* **2013**, *21*, 463–476.

4. García-Haro, F.J.; Camacho, F.; Meliá, J. *Vegetation Parameters Validation Report (VEGA VR)*, SAF/LAND/UV/VR VEGA/2.1; Available online: http://landsaf.meteo.pt (accessed on 19 June 2014).

5. Yang, W.; Huang, D.; Tan, B.; Stroeve, J.C.; Shabanov, N.V.; Knyazikhin, Y.; Nemani, R.R.; Myneni, R.B. Analysis of leaf area index and fraction of PAR absorbed by vegetation products from the Terra MODIS sensor: 2000–2005. *IEEE Trans. Geosci. Remote Sens* **2006**, *44*, 1829–1842.

6. Goward, S.; Markham, B.; Dye, D. Normalized difference vegetation index measurements from the advanced very high resolution radiometer. *Remote Sens. Environ* **1991**, *35*, 257–277.

7. Holben, B.N. Characteristics of maximum-value composite images from temporal AVHRR data. *Int. J. Remote Sens***1986**, *7*, 1417–1434.

8. Kobayashi, H.; Dye, D.G. Atmospheric conditions for monitoring the long-term vegetation dynamics in the Amazon using normalized difference vegetation index. *Remote Sens. Environ* **2005**, *97*, 519–525.

9. Gutman, G. Vegetation indices from AVHRR: An update and future prospects. *Remote Sens. Environ* **1991**, *35*, 121–136.

10. Roerink, G.J.; Menenti, M.; Verhoef, W. Reconstructing cloud free NDVI composites using Fourier analysis of time series. *Int. J. Remote Sens* **2000**, *21*, 1911–1917.

11. Fensholt, R.; Rasmussen, K.; Nielsen, T.T.; Mbow, C. Evaluation of Earth observation based long term vegetation trends—Intercomparing NDVI time series trend analysis consistency of Sahel from AVHRR GIMMS, Terra MODIS and SPOT VGT data. *Remote Sens. Environ* **2009**, *113*, 1886–1898.

12. Gao, F.; Morisette, J.; Wolfe, R.E.; Ederer, G.; Pedelty, J.; Masuoka, E.J.; Myneni, R.; Tan, B.; Nightingale, J.M. An algorithm to produce temporally and spatially continuous MODIS LAI time series. *IEEE Geosci. Remote Sens. Lett* **2008**,*5*, 60–64.

13. Rossi, R.E.; Dungan, J.L.; Beck, L.R. Kriging in the shadows: Geostatistical interpolation for remote sensing. *Remote Sens. Environ* **1994**, *49*, 32–40.

14. De Oliveira, J.C.; Epiphanio, J.C.N.; Rennó, C.D. Window regression: A spatial-temporal analysis to estimate pixels classified as low-quality in MODIS NDVI time series. *Remote Sens* **2014**, *6*, 3123–3142.

15. Potter, C.; Tan, P.N.; Steinbach, M.; Klosster, S.; Kumar, V.; Myneni, R.; Genovese, V. Major disturbance events in terrestrial ecosystems detected using global satellite data sets. *Glob. Chang. Biol* **2003**, *9*, 1005–1021.

16. Kandasamy, S.; Baret, F.; Verger, A.; Neveux, P.; Weiss, M. A comparison of methods for smoothing and gap filling time series of remote sensing observations: Application to MODIS LAI products. *Biogeosciences* **2013**, *9*, 17053–17097.

17. Jönsson, P.; Eklundh, L. Seasonality extraction by function fitting to time-series of satellite sensor data. *IEEE Trans. Geosci. Remote Sens* **2002**, *40*, 1824–1832.

18. Julien, Y.; Sobrino, J.A. Comparison of cloud-reconstruction methods for time series of composite NDVI data. *Remote Sens. Environ* **2010**, *114*, 618–625.

19. Chen, J.; Per, J.; Masayuki, T.; Gu, Z.H.; Bunkei, M.; Lars, E. A simple method for reconstructing a high-quality NDVI time-series data set based on the Savitzky-Golay filter. *Remote Sens. Environ* **2004**, *91*, 332–344.

20. Iwaniec, J.; Lisowski, W.; Uhl, T. Nonparametric approach to improvement of quality of modal parameters estimation.*J. Theor. Appl. Mech* **2005**, *43*, 327–344.

21. Cleveland, W.S.; Devlin, S.J. Locally-Weighted Regression: An Approach to Regression Analysis by Local Fitting. *J. Am. Stat. Assoc* **1988**, *83*, 596–610.

22. Atzberger, C.; Eilers, P.H. Evaluating the effectiveness of smoothing algorithms in the absence of ground reference measurements. *Int. J. Remote Sens* **2011**, *32*, 3689–3709.

23. Verger, A.; Baret, F.; Weiss, M.; Kandasamy, S.; Vermote, E.F. The CACAO method for smoothing, gap filling, and characterizing seasonal anomalies in satellite time series. *IEEE Trans. Geosci. Remote Sens* **2013**, *51*, 1963–1972.

24. Hutchinson, M.F.; De Hoog, F.R. Smoothing noisy data with spline functions. *Numer. Math* **1985**, *47*, 99–106.

25. Jönsson, P.; Eklundh, L. TIMESAT—A program for analyzing time series of satellite sensor data. *Comput. Geosci* **2004**,*30*, 833–845.

26. Stöckli, R.; Vidale, P.L. European plant phenology and climate as seen in a 20-year AVHRR land-surface parameter dataset. *Int. J. Remote Sens* **2004**, *25*, 3303–3330.

27. Beck, P.S.; Atzberger, C.; Høgda, K.A.; Johansen, B.; Skidmore, A.K. Improved monitoring of vegetation dynamics at very high latitudes: A new method using MODIS NDVI. *Remote Sens. Environ* **2006**, *100*, 321–334.

28. Bradley, B.A.; Jacob, R.W.; Hermance, J.F.; Mustard, J.F. A curve fitting procedure to derive inter-annual phenologies from time series of noisy satellite NDVI data. *Remote Sens. Environ* **2007**, *106*, 137–145.

29. Azzali, S.; Menenti, M. Mapping vegetation–soil–climate complexes in southern Africa using temporal Fourier analysis of NOAA-AVHRR NDVI data. *Int. J. Remote Sens* **2000**, *21*, 973–996.

30. Sakamoto, T.; Wardlow, B.D.; Gitelson, A.A.; Verma, S.B.; Suyker, A.E.; Arkebauer, T.J. A two-step filtering approach for detecting maize and

soybean phenology with time-series MODIS data. *Remote Sens. Environ* **2010**, *114*, 2146–2159.

31. Martínez, B.; Gilabert, M.A. Vegetation dynamics from NDVI time series analysis using the wavelet transform. *Remote Sens. Environ* **2009**, *113*, 1823–1842.

32. Tan, B.; Morisette, J.T.; Wolfe, R.E.; Gao, G.; Ederer, G.; Nightingale, J.; Pedelty, J. An enhanced TIMESAT algorithm for estimating vegetation phenology metrics from MODIS data. *IEEE J. Sel. Top. Appl. Earth Obs. Remote Sens* **2011**, *4*, 361–371.

33. Strahler, A.; Muchoney, D.; Borak, J.; Gao, F.; Friedl, M.; Gopal, S.; Hodges, J.; Lambin, E.; McIver, D.; Moody, A.; Schaaf, C.; Woodcock, C. MODIS Land Cover Product, Algorithm Theoretical Basis Document (ATBD), Version 5.0. Available online: http://modis.gsfc.nasa.gov/data/atbd/ (accessed on 19 June 2014).

34. Hird, J.N.; McDermid, G.J. Noise reduction of NDVI time series: An empirical comparison of selected techniques. *Remote Sens. Environ* **2009**, *113*, 248–258.

35. Musial, J.P.; Verstraete, M.M.; Gobron, N. Technical Note: Comparing the effectiveness of recent algorithms to fill and smooth incomplete and noisy time series. *Atmos. Chem. Phys* **2011**, *11*, 7905–7923.

36. Geng, L.; Ma, M.; Wang, X.; Yu, W.; Jia, S.; Wang, H. Comparison of eight techniques for reconstructing multi-satellite sensor time-series NDVI data sets in the Heihe river basin, China. *Remote Sens* **2014**, *6*, 2024–2049.

37. Atkinson, P.M.; Jeganathan, C.; Dash, J.; Atzberger, C. Inter-comparison of four models for smoothing satellite sensor time-series data to estimate vegetation phenology. *Remote Sens. Environ* **2012**, *123*, 400–417.

38. Yuan, H.; Dai, Y.; Xiao, Z.; Ji, D.; Shangguan, W. Reprocessing the MODIS Leaf Area Index products for land surface and climate modeling. *Remote Sens. Environ* **2011**, *115*, 1171–1187.

39. Roujean, J.L.; Bréon, F.M. Estimating PAR absorbed by vegetation from bidirectional reflectance measurements. *Remote Sens. Environ* **1995**, *51*, 375–384.

40. Roujean, J.L.; Lacaze, R. Global mapping of vegetation parameters from POLDER multiangular measurements for studies of surface-atmosphere interactions: A pragmatic method and its validation. *J. Geophys. Res.: Atmos* **2002**, *107*, 1984–2012.

41. Verger, A.; Camacho-de Coca, F.; Meliá, J. Inter-comparison of algorithms for retrieving operationally vegetation parameters at global scale: Assessment over Europe along 2003. In Proceedings of the 2nd Recent

Advances in Quantitative Remote Sensing, Torrent, Spain, 25–29 September 2006; pp. 909–914,

42. Camacho, F. Evaluation of the Land-SAF FAPAR Prototype along One Year of MSG BRDF Data: Algorithm, Product Description, and Intercomparison against Equivalent Satellite Products and Ground-Truth. Available online: http://landsaf.meteo.pt/documentsView.jsp (accessed on 20 June 2014).

43. Schaaf, C.B.; Gao, F.; Strahler, A.H.; Lucht, W.; Li, X.; Tsang, T.; Strugnell, N.C.; Zhang, X.; Jin, Y.; Muller, J.-P; et al. First operational BRDF, albedo nadir reflectance products from MODIS. *Remote Sens. Environ* **2002**, *83*, 135–148.

44. Madsen, K.; Nielsen, H.B.; Tingleff, O. *Methods for Non-Linear Least Squares Problems*; Informatics and Mathematical Modeling, Technical University of Denmark: Lyngby, Denmark, 2004.

45. Madden, H.H. Comments on the Savitzky-Golay convolution method for least-squares-fit smoothing and differentiation of digital data. *Anal. Chem* **1978**, *50*, 1383–1386.

46. Craven, P.; Wahba, G. Smoothing noisy data with spline functions. *Numer. Math* **1978**, *31*, 377–403.

47. Eklundh, L.; Jönsson, P. *TIMESAT 3.1 Software Manual*; Lund University: Lund, Sweden, 2012.

48. Press, W.H.; Teukolsky, S.A.; Vetterling, W.T.; Flannery, B.P. *Numerical Recipes: The Art of Scientific Computing*, 3rd ed.; Cambridge University Press: Cambridge, UK, 2007.

49. Pérez-Hoyos, A.; García-Haro, F.J.; San-Miguel-Ayanz, J. A methodology to generate a synergetic land-cover map by fusion of different land-cover products. *Int. J. Appl. Earth Obs. Geoinf* **2012**, *19*, 72–87.

50. Michishita, R.; Jin, Z.; Chen, J.; Xu, B. Empirical comparison of noise reduction techniques for NDVI time-series based on a new measure. *ISPRS J. Photogramm. Remote Sen* **2014**, *91*, 17–28.

Chapter 12

VDTA-Based Wave Active Filter

Harshvardhan Singh[1], Kunal Arora[1], Dinesh Prasad[2*]

[1]*Department of Applied Sciences and Humanities, Faculty of Engineering and Technology, Jamia Millia Islamia, New Delhi, India*

[2]*Department of Electronics and Communication Engineering, Faculty of Engineering and Technology, Jamia Millia Islamia, New Delhi, India*

ABSTRACT

In this paper we present a wave active filter based on Voltage Differencing Transconductance Amplifiers (VDTAs). The synthesis of active filters basically based on processing of wave quantities. The wave method is presented for basic building blocks of active filters i.e. a series inductor and parallel capacitor through which realization of various active circuits is made by appropriate connections. The proposed wave active filter is verified by realizing a 4th order low pass Butterworth filter using SPICE simulation with 0.18 μm TSMC CMOS technology parameters.

Keywords: Voltage Differencing Transconductance Amplifier, Wave Active Filter, Voltage-Mode Filter

1. INTRODUCTION

The high order active filters can be realized by imitating the behavior of elements of LC ladder prototype filters and the approach for the designing of these filters has been already discussed in the literature [1] -[6] and the references cited therein. The purpose of the method was to derive active filters based on scattering parameters. Synthesis of active filters is based on the use of wave quantities, hence the scattering matrix will play an important role in the concept, as already discussed in the reference [1] and [2] . Wave active filters using various Active building blocks (ABB) are available in the literature [3] -[6] such as Current Feedback Operational Amplifiers [3] , Differential Voltage Current Conveyor Transconductance Amplifier [4] , Current Controlled Differential Difference Current Conveyor Transconductance Amplifier [5] and Operational Trans-Resistance Amplifier [6] .

This paper presents the realization of wave active filter using a recently introduced ABB VDTA. The advantages and usefulness of VDTA are discussed in references [7] -[12] .

Wave equivalent is developed for an inductor in series branch and for a capacitor in parallel branch using VDTAs. The workability of 4th order low pass Butterworth filter is thus verified through SPICE simulation using 0.18 μm TSMC CMOS technology parameters.

2. VDTA DESCRIPTION

The VDTA is a recently introduced active element which has two voltage inputs and two kinds of current output. The symbol of VDTA is shown in Figure 1 and its CMOS implementation is shown in Figure 2 [7] , where the input terminals are denoted as V_P and V_N and output terminals are Z, X^+ and X^-. The terminal relationship of VDTA can be described by the following set of equations:

$$\begin{bmatrix} I_Z \\ I_{X^+} \\ I_{X^-} \end{bmatrix} = \begin{bmatrix} g_{m_1} & -g_{m_1} & 0 \\ 0 & 0 & g_{m_2} \\ 0 & 0 & -g_{m_2} \end{bmatrix} \begin{bmatrix} V_{V_P} \\ V_{V_N} \\ V_Z \end{bmatrix}$$

(1)

The CMOS realization of VDTA is shown in Figure 2 with,

$$g_{m_1} = (g_3 + g_4)/2$$

(2a)

$$g_{m_2} = (g_5 + g_8)/2 \quad \text{or} \quad g_{m_2} = (g_6 + g_7)/2$$

(2b)

where g_i is the called as the Transconductance value of ith transistor defined by

$$g_i = \sqrt{I_{Bi} \cdot \mu_i \cdot C_{ox} \cdot \frac{W}{L}}$$

(3)

3. BASIC WAVE EQUIVALENT USING VDTA

3.1. Basic Wave Equivalent

For defining the practicality of the filter the wave method is used and defined by the scattering matrix S. The incident and reflected voltage waves are illustrated as A_i and B_i respectively for two port network of Figure 3 and are related by the following relation:

$$A_i = V_i + I_i R_i \, , \, B_i = V_i - I_i R_i$$

(4)

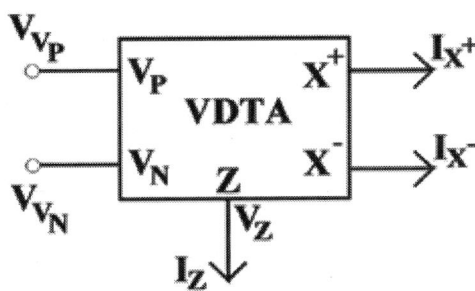

Figure 1. Symbol of VDTA.

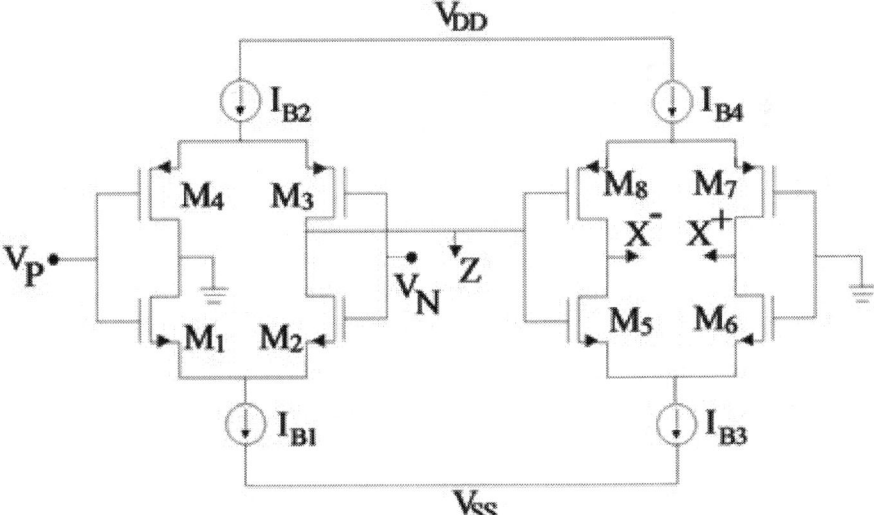

Figure 2. CMOS implementation of VDTA [7] .

Figure 3. Incident waves A1, A2 and reflected waves B1, B2 for two port network.

Equation (4) can be expressed in terms of scattering matrix S as

$$\begin{bmatrix} B_1 \\ B_2 \end{bmatrix} = S \begin{bmatrix} A_1 \\ A_2 \end{bmatrix}$$

(5)

The series inductor L and parallel capacitor C can be described in terms of scattering parameter given by Equation (6) for L and C respectively.

$$S = \frac{1}{1+s\tau}\begin{bmatrix} s\tau & 1 \\ 1 & s\tau \end{bmatrix} \quad \text{and} \quad S = \frac{1}{1+s\tau}\begin{bmatrix} -s\tau & 1 \\ 1 & -s\tau \end{bmatrix}$$

(6)

By going through the concept of wave filtering using the scattering parameter description, the incident waves (A_i) and the reflected waves (B_i) of an inductor (L) in series branch, are expressed as Equations (7) and (8).

$$B_1 = A_1 - \frac{1}{1+s\tau}(A_1 - A_2)$$

(7)

$$B_2 = A_2 + \frac{1}{1+s\tau}(A_1 - A_2)$$

(8)

where $\tau = L/2R$ is time constant and R is the characteristic resistance assigned at each port named port resistance [1] [2] . Similarly for a capacitor (C) in shunt branch the equations are (9) and (10) where $\tau = RC/2$.

$$-B_1 = A_1 - \frac{1}{1+s\tau}(A_1 - A_2)$$

(9)

$$B_2 = -A_2 + \frac{1}{1+s\tau}(A_1 + A_2)$$

(10)

Equations (7), (8), (9) and (10), can be realized by the use of following processes: 1) Lossy Integration-Subtraction, 2) Subtraction and 3) Summation.

3.2. Lossy Integration-Subtraction

A Lossy integration-subtraction configuration is shown in Figure 4. It uses a single VDTA, a parallel combination of resistor R_2 and capacitor C_d at output terminal X^+ and also a grounded resistor R_1 at output terminal Z.

The input-output relationship is given by the following equation:

$$V_O = \frac{1}{1+s\tau}\left(V_{in1} - V_{in2}\right)$$

(11)

where the realized time constant $\tau = R_2 C_d$, $g_{m1}R_1 = 1$ and $g_{m2}R_2 = 1$. On comparing Equation (11) with equations (7) and (8), it is accomplished that the following condition must be fulfilled: $R_2 C_d = L/2R$. Considering, port resistance $R = R_2$, the value of capacitor in wave active realization is given by:

$$C_d = L/2R^2$$

(12)

Subtraction:

To implement the subtraction operation using VDTA, the configuration is depicted in Figure 5.

$$V_O = \left(V_{in1} - V_{in2}\right) \text{ with } g_{m1}R_1 = 1 \text{ and } g_{m2}R_2 = 1$$

(13)

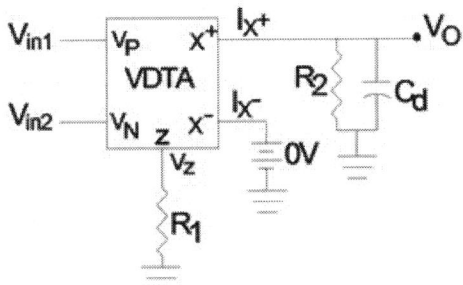

Figure 4. Lossy integration-subtraction using VDTA.

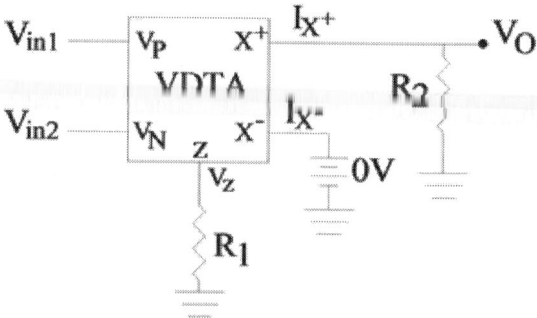

Figure 5. Subtraction using VDTA.

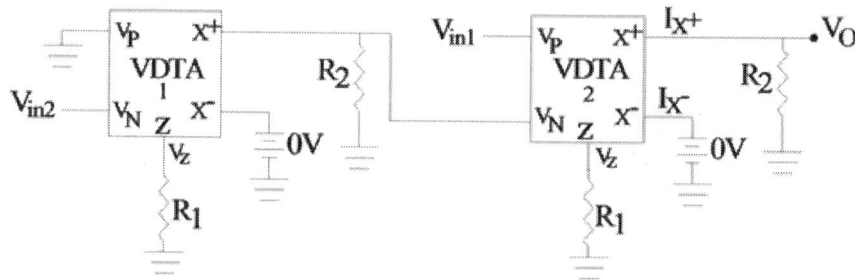

Figure 6. Summation using VDTA.

Summation:

A VDTA based summation configuration is shown in Figure 6. It consists of two VDTAs. The first VDTA reverses the input V_{in2} which is then subtracted from input V_{in1} by second VDTA to give,

$$V_O = \left(V_{in1} + V_{in2}\right) \text{ with } g_{m1}R_1 = 1 \text{ and } g_{m2}R_2 = 1 \tag{14}$$

Using the blocks in Figure 4, Figure 5 and Figure 6, the resultant wave equivalent of an inductor in seriesbranch as given by Equations (7) and (8) is shown in Figure 7 and its symbolic representation is shown in Figure 8.

Similarly, the resultant wave equivalent of a capacitor in shunt-branch as given by Equations (9) and (10) is shown in Figure 9 and its symbolic representation is shown in Figure 10.

4. COMPLETE SET OF WAVE EQUIVALENTS

According to the wave method, the wave flow diagrams that could be employed for designing active filters are summarized in Table 1. The required inversion blocks could be obtained by employing the subtraction block in Figure 5 with the condition that $V_{in1} = 0$.

To accomplish the construction of whole filter circuit, the main points are: port resistances are assumed to be equal and the cross-cascade connection of the incident and reflected waves is applied because the incident wave at each port equals the reflected wave at the foregoing port [1] -[6] . Wave equivalents are substituted in place of individual capacitors and inductors and the complete structure is then achieved by cascading the respective wave equivalents.

5. SIMULATION RESULTS

Simulations are performed by using SPICE program with TSMC CMOS 0.18 μm process parameters. The aspect ratios of various transistors used are given in Table 2. Supply voltages are taken as $V_{DD} = -V_{SS} = 0.9$ V and the transconductances of VDTA were controlled by bias currents $I_{B1} = I_{B2} = I_{B3} = I_{B4} = 513.36$ uA. Thus, the transconductance were found to be $g_{m1} = g_{m2} = 1$ mA/V. For verification of the suggested method defined in Sections 3 and 4, a 4th order low pass Butterworth filter (Figure 11) has been taken for experiment. The component values used are $R_S = R_L = 1$ KΩ, $L_1 = 0.2437$ mH, $L_2 = 0.5884$ mH, $C_1 = 0.5884$ nF, $C_2 = 0.2437$ nF.

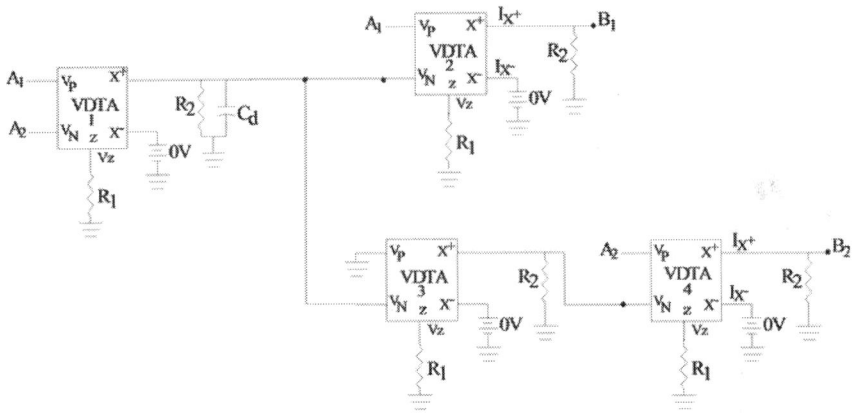

Figure 7. VDTA based wave equivalent of an inductor in series branch.

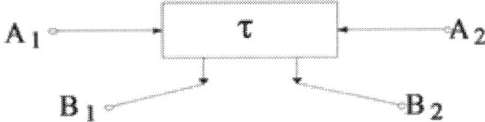

Figure 8. Symbolic representation wave equivalent of series inductor.

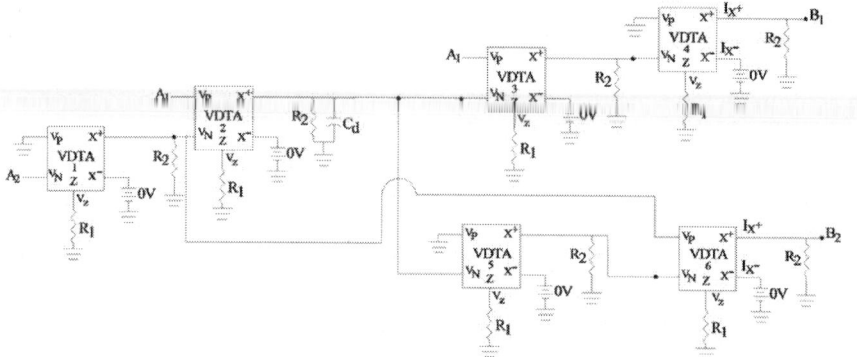

Figure 9. VDTA based wave equivalent of a capacitor in parallel branch.

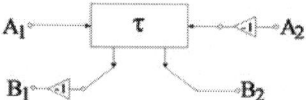

Figure 10. Symbolic representation of wave equivalent of parallel capacitor.

The filter circuit is implemented using the wave equivalents of series inductor and parallel capacitor. The theoretically predicted value of cutoff frequency and also measured by the response of 4th order low pass Butterworth filter shown in Figure 12 is 511.48 KHz. The resistors R_1, R_2 are chosen to be 1 kΩ according to $g_{m1}R_1 = g_{m2}R_2 = 1$. The values of capacitor C_d for wave equivalent of series inductors (L_1, L_2) are 0.12185 nF, 0.294 nF and for wave equivalent of shunt capacitors (C_1, C_2) are 0.2942 nF, 0.12185 nF. The complete structure obtained by cascading the wave equivalents is shown in Figure 13 and has been simulated using VDTA based wave equivalents. Figure 14 and Figure 15 display the simulated filter responses for 4th order low pass (V_o) and its complementary high pass (V_{oc}) respectively. The measured cutoff frequency of the filter was 503.52 KHz.

Table 1. Wave equivalent of elementary two port consisting of single element.

Elementary two port	Wave equivalent	Time constant	Capacitor value for VDTA based wave equivalent
L ⎓ R ⎓ R	A₁—τ—A₂ B₁—B₂	$\tau = \dfrac{L}{2R}$	$C_d = L/2R^2$
C ⎓ R ⎓ R	A₁—τ—A₂ B₁—B₂	$\tau = 2RC$	$C_d = 2C$
L ⎓ R ⎓ R	A₁—τ—A₂ B₁—B₂	$\tau = \dfrac{2L}{R}$	$C_d = 2L/R^2$
C ⎓ R ⎓ R	A₁—τ—A₂ B₁—B₂	$\tau = \dfrac{RC}{2}$	$C_d = C/2$

Table 2. Transistors aspect ratios for VDTA.

Transistors	W (μm)	L (μm)
M_1, M_2, M_5, M_6	3.6	0.36
M_3, M_4, M_7, M_8	16.64	0.36

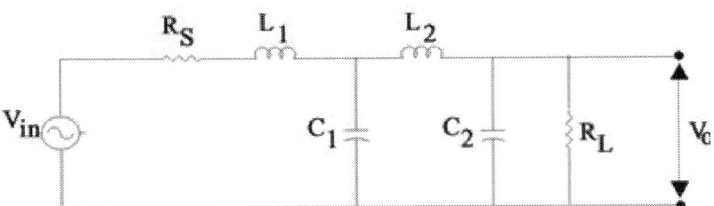

Figure 11. 4th order low pass butterworth filter.

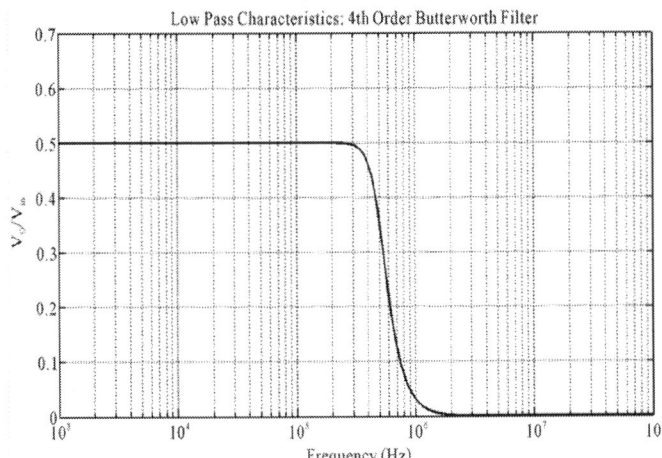

Figure 12. Frequency response of 4th order low pass butterworth filter.

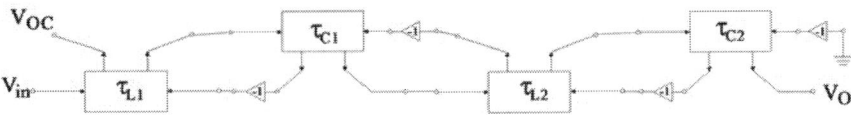

Figure 13. Wave equivalent of 4th order butterworth filter.

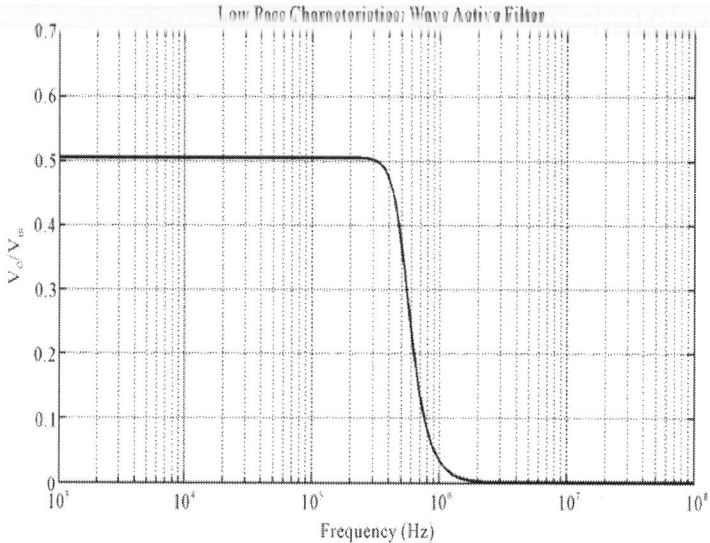

Figure 14. Frequency response of 4th order low pass filter: wave active filter.

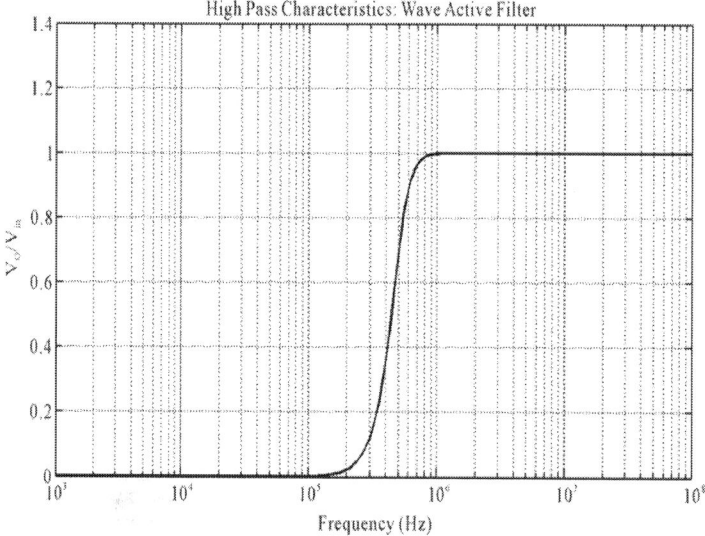

Figure 15. Frequency response of 4th order complementary high pass filter: wave active filter.

6. CONCLUSION

A new wave active filter is presented using recently introduced ABB VDTA. VDTAs are used to design lossy integration-subtraction, subtraction and summation blocks, which are the main steps in realizing the wave active filter. The wave method is verified by realizing the 4th order low pass and high pass responses. The proposed wave filter may be designed by using other ABBs which requires lesser power consumption than VDTAs. SPICE simulation results thus confirm the operation of wave active filter with 0.18 μm TSMC CMOS technology parameters.

REFERENCES

1. Wupper, H. and Meerkotter, K. (1975) New Active Filter Synthesis Based on Scattering Parameters. IEEE Transactions on Circuits and Systems, 22, 594-602.
2. Haritantis, I., Constantinides, A. and Deliyannis, T. (1976) Wave Active Filters. IEE Proceedings, 123, 676-682.
3. Koukiou, G. and Psychalinos, C. (2010) Modular Filter Structures Using Current Feedback Operational Amplifiers. Radioengineering, 19, 662-666.
4. Pandey, N. and Kumar, P. (2011) Differential Voltage Current Conveyer Transconductance Amplifier Based Wave Active Filter. Journal of Electronic Devices, 10, 429-432.
5. Pandey, N., Kumar, P. and Choudhary, J. (2013) Current Controlled Differential Difference Current Conveyor Transconductance Amplifier and Its Applications. ISRN Electronics, 2013, Article ID: 968749.
6. Bothra, M., Pandey, R., Pandey, N. and Paul, S.K. (2013) Operational Trans-Resistance Amplifier Based Tunable Wave Active Filter. Radioengineering, 22, 159-166.
7. Yesil, A., Kacar, F. and Kuntman, H. (2011) New Simple CMOS Realization of Voltage Differencing Transconductance Amplifier and Its RF Filter Application. Radioengineering, 20, 632-637.
8. Prasad, D. and Bhaskar, D.R. (2012) Electronically Controllable Explicit Current Output Sinusoidal Oscillator Employing Single VDTA. ISRN Electronics (USA), 2012, Article ID: 382560.
9. Prasad, D. and Bhaskar, D.R. (2012) Grounded and Floating Inductance Simulation Circuits Using VDTAs. Circuits and Systems (USA), 3, 342-347.
10. Prasad, D., Bhaskar, D.R. and Srivatava, M. (2013) Universal Current-Mode Biquad Filter Using a VDTA. Circuits and Systems (USA), 4, 29-33.

11. Prasad, D., Srivatoya, M. and Bhaskar, D.R. (2013) Electronically Controllable Fully-Uncoupled Explicit CurrentMode Quadrature Oscillator Using VDTAs and Grounded Capacitors. Circuits and Systems (USA), 4, 169-172.

Index

A

accelerometers, 229, 230, 246

Acoustic Echo, 14

Algorithm, 26, 27, 35, 36, 93, 106, 108, 115, 118, 162, 168, 169, 237, 238, 242, 243, 277, 278

attitude control, 230

B

Ballistocardiography, 179, 184

C

combination forecast, 26, 27, 31, 34, 35, 43, 44, 46

D

data fusion, 61, 62, 63, 73, 74, 75, 83, 84, 86, 232, 246

G

gas turbine, 61, 62, 63, 64, 65, 66, 67, 68, 72, 74, 75, 77, 83, 84, 85

H

Hammerstein adaptive filtering, 87, 88, 89, 93, 100, 101, 102, 105

J

Jammer suppression, 6

L

Linear predictor, 4

load forecasting, 25, 26, 27, 49, 50, 51, 55, 56, 57, 58

M

Markov chain, 26, 32, 44, 50, 56, 58

Methodology, 137

N

nonlinear system identification, 87, 88

O

Ocular Artifacts, 19, 136, 160, 162

P

Photoplethysmography, 111, 120, 181

probability density truncation, 61

R

robustness, 25, 26, 50, 61, 87, 105, 128

S

Spectrogram, 10

System identification, 2

V

Voltage-Mode Filter, 279

W

Wave Active Filter, 279, 289

wavelet transform, 55, 61, 71, 79, 86, 178, 182, 248, 277